SDI: A View from Europe

SDI

A View From Europe

ROBERT C. HUGHES

NATIONAL DEFENSE UNIVERSITY PRESS
Fort Lesley J. McNair
Washington, DC

National Defense University Press Publications. To increase general knowledge and inform discussion, NDU Press publishes books on subjects relating to US national security. Each year in this effort, The National Defense University, through the Institute for National Strategic Studies, hosts about two dozen Senior Fellows who engage in original research on national security issues. NDU Press publishes the best of this research. In addition, the Press publishes especially timely or distinguished writing on national security from authors outside the University, new editions of out-of-print defense classics, and books based on conferences concerning national security affairs.

The Far Side cartoon on page 181 is copyrighted and reprinted by permission. The remainder of this publication is not copyrighted and may be quoted or reprinted without permission. Please give full publication credit.

Opinions, conclusions, and recommendations expressed or implied within are solely those of the author and do not necessarily represent the views of The National Defense University, the Department of Defense, or any other Government agency. Cleared for public release; distribution unlimited.

NDU Press publications are sold by the US Government Printing Office. For ordering information, call (202) 783-3238, or write to: Superintendent of Documents, US Government Printing Office, Washington, DC 20402.

Primary Sources Research, Chevy Chase, Maryland, indexed this publication under contract DAHC-32-92M0247.

Library of Congress Cataloging-in-Publication Data
Hughes, Robert C.
    SDI: a view from Europe / Robert C. Hughes.
        p. cm.
    Includes index.
    $8.00 (est.)
    1. Strategic Defense Initiative. 2. Ballistic missile defenses—Europe. 3. North Atlantic Treaty Organization—Armed Forces. 4. United States—Military relations—Europe. 5. Europe—Military relations—United States. I. Title.
UG743.H84 1990
358'.1754—dc20   90-37440
                                                                                   CIP

First printing, December 1990

FOR
SANDE, BRIDGET, TIMOTHY, AND BRENDAN

Contents

Foreword   *xiii*

Preface   *xv*

Chapter 1.	To Avenge or to Defend?	3
	Point of Departure	4
	Objectives of SDI	11
	Nitze Criteria	16
	SDI: Contrarieties	17
	Multiple Perspectives of the Un-Definition of SDI	27
2.	European Candling of SDI	33
	Criticism from Europe	33
	Phases of Reaction	35
	Major Concerns Rarely Examined	60
	Deterrence Strategy Revisited	61
3.	SDI at European Capitals and NATO	73
	Managing SDI in NATO Europe	73
	Lord Carrington and the NATO Staffs	76
	Extended Deterrence versus Extended Defense	84
	Alliance Participation	87

		Reflections on Issues Raised by the "Big Four"	94
		Issues in Reserve	119
CHAPTER	4.	IMPLICATIONS FOR NATO'S STRATEGY	123
		NATO's Strategic Concept	126
		Effects on NATO's Strategic Concept	129
		Deterrence versus Defense	136
		Responses to Strategic Defenses	139
		Strategic Stability	143
		Stability: A Prism of Concepts	145
	5.	POTENTIAL FOR A STABLE TRANSITION	161
		Central Strategic Concept	161
		Considerations in Any Transition	167
		The Purists and Absolutists	182
		Discriminators: Old Business with a New Mix	185
		Exploiters and Other Piggybackers	187
		Apollonian and Dionysian Arms Control	190
		Bargaining Chip or Chip on the Shoulder?	191
	6.	SDI'S BEARING ON NATO	195
		Misologists Need Not Apply	196
		Concluding Thoughts and Recommendations	201

APPENDICES	A. SUMMARY OF STATUS OF ALLIED CONTRACTS	215
	B. FY 85-90 SDI FUNDING LEVELS	216

Notes   217

Glossary   239

Index   247

About the Author   255

Foreword

Since its announcement in 1983, the Strategic Defense Initiative has been the subject of considerable controversy. The range of opinion among commentators in the United States is well documented, but the views of Western Europeans about SDI are less well known to American readers.

In *SDI: A View from Europe*, Colonel Robert C. Hughes, USAF, explains the West European responses to this American initiative. He discusses the major issues raised by Europeans, among them the wisdom of trying to negate the threat of ballistic missiles by constructing a sophisticated interception system. He analyzes the argument presented by some Europeans that deterrence and stability might be diminished by a new system promising to limit damage by destroying incoming ballistic missiles. And he examines the often voiced European objection that SDI, by promising so strong a defense, might encourage warfighting rather than maintain deterrence.

In addition to presenting a comprehensive view of European attitudes toward the Strategic Defense Initiative, Colonel Hughes also illuminates the new threat posed by ballistic missiles in the hands of the score or more nations that will possess them by the mid-1990s. *SDI: A View from Europe* concludes by discussing the possibility that the Soviets in particular may find such defenses to be in their own interest.

J. A. BALDWIN
VICE ADMIRAL, US NAVY
PRESIDENT, NATIONAL DEFENSE UNIVERSITY

Preface

In an informal exchange at NATO Headquarters in the early spring of 1987, a few of the more experienced and less anxious ambassadors set forth views on the Strategic Defense Initiative (SDI) that have proven at once prudent, resilient, and venturesome. The brief they argued included the following points: do not worry, nothing is about to happen; NATO strategy will not be changed anytime soon because of SDI; especially with science moving ahead so fast, the United States should continue looking into the technologies and, along with the allies, into the strategy issues; nothing should be done that the Soviet Union would misunderstand and perceive as a threat; the superpowers should keep the ABM Treaty intact until there is certainty that defensive forces can strengthen stability as some American strategists assert; and, finally, the questions SDI raises will continue to be important because they strike at the core of NATO's strategic concept for deterrence and defense. Such issues centered on damage limitation, the improved survivability of nuclear forces for retaliation, and the devaluation of the military and political worth of ballistic missiles.

Most in attendance that March day believed that ambassadors several times removed from the current permanent representatives would still be discussing SDI and the relationship between the offense and the defense. Although "defensive deterrence" might some day supplement if not supplant deterrence based primarily on offensive forces, such an eventuality was not at all certain. Political, economic, and social changes in the balance of power would be at least as important as changes in the military equation. In brief, there would be time for assessments of the implications of SDI for strategy before any decision to deploy a "splendid" system of defenses.

Even with all the cautions in place, ambassadors agreed that the important thing was not to stop asking questions of scientists, strategists, and arms controllers about the permanence, relevance, and character of deterrence and the undergirding nuclear forces. Questions about the efficacy of active defenses against ballistic missiles, the potential for transition to a deterrence built on a balance of decreased offense and increased defense, the focus on the weapons to be destroyed rather than on what is to be protected, the projected proliferation of ballistic missile technologies and the fielding of short- to intermediate-range systems in a score or more nations by the early 1990s, the push and pull of SDI on arms control negotiations and on the ABM Treaty, the "sanctuary" of space free of stationed weapons—these were some of the principal issues ambassadors dealt with as they struggled to know more about SDI and its contributions, if any, to stability and predictability.

Although none of the ambassadors showed the slightest prescience about the dazzling changes that were to occur in 1989 and 1990 in Europe, each of them did have dreams for better relations with the Soviet Union and the East Central European nations. SDI threatened that vision. Several nations thought they detected in SDI the necessary scuttling of chances for major arms reductions, improved dialogue, and the beneficial entanglements of increased commerce with the Warsaw Pact nations. Political changes in Europe and shifts in the military balance over the past two years, along with arms control expectations, have had salutary effects in lessening tensions. At the same time, however, recent aggression in the Persian Gulf, coupled with proliferation of ballistic missiles across the globe, has increased the urgency for developing defenses against ballistic and cruise missiles.

From the fall of 1984 to the summer of 1987, I was privileged to serve in the Defense Plans Division at the United States Mission to NATO at Evere. During those years, I worked on SDI issues for Ambassador David M. Abshire (and during a few months in 1987 for Ambassador Alton J. Keel) as well as for the then Defense Advisor Dr. Laurence J. Legere. In the fall of 1984, Secretary Shultz and Assistant Secretary Burt had asked Ambassador Abshire to be the administration's "point man" in Europe on SDI. In that role, Ambassador Abshire established a working group on which I served, involving State, Defense, and USIA elements within the Mission; the group's charter was to stay in front of the technology, strategy, and arms control issues Europeans were raising in regard to SDI.

Those who know Ambassador Abshire will appreciate the whirlwind of concepts, initiatives, and "actions" he is capable of inventing during any single week, day, or hour. Those who know Dr. Legere will appreciate the herculean efforts he was capable of in translating concepts into practicable proposals and programs. He knew how NATO worked, and he could make it hum when he was convinced of the merits of any proposals. I deeply admire both of my former bosses at NATO and am grateful they gave me great latitude in working with allies on the policy and strategy dimensions of SDI. In fact, the origins of this book lay with tasks each of them separately gave me to explore the case that European nations, particularly NATO allies, were making about SDI.

The cover the editor has chosen for this text captures just how different the European view of SDI is from the view offered to Europe by senior officials in the Reagan and Bush administrations. The European view is distinctly earth-bound, with ominous storm clouds and promises of clearing. In my three years at Brussels working on NATO's defense plans and policies, I heard the term *space* as a potential environment for combat used less than a handful of times. The military establishments knew better, but the political establishments found it convenient to ignore space issues or, when necessary, to handle them quietly. Europe's principal concern remained the land, the air above it, and the adjacent seas; that was the perspective from which many Europeans saw SDI. From that viewpoint, SDI had not many admirers.

I am grateful to the many friends and colleagues who read drafts of this book at various stages and who tolerated long, I hope not enervating, discussions with me as I tried out ideas on them. While I risk not including all whom I should, I want to mention a number whose comments I found particularly insightful and instructive. Many thanks to John Reichart, Mike Moodie, and Steve Sturm—day-and-night laborers one and all at the US Mission to NATO.

I also warmly acknowledge the encouragement and expertise of fellow faculty members and students at the National War College who gave me the benefit of their reflections on individual nation's reactions to SDI, as well as on matters of strategy, nuclear forces, and deterrence theory—my thanks to Bob Beecroft, Bob Gallucci, Ilana Kass, Tom

Keaney, Al Pierce, Roy Stafford, and Steve Szabo. Within the wider University community, I thank Tom Julian for his generous assistance as a reader. I also am indebted to many within the Strategic Defense Initiative Organization, especially Dave Martin who showed the endurance of a professional and the patience of a friend in reviewing many drafts; as one daily engaged with European responses to the SDI program, his help was of great merit. I also single out B. A. Myers from the US Mission as one of the nicest people on the earth and one who pushed me to think through the European approaches to this American initiative. My gratitude also goes to Jack Swartwood, a long-time friend, who gave excellent technical guidance in the preparation of the text.

Finally, special thanks to Dr. Fred Kiley, director of the National Defense University Press. He was my first boss some twenty-years ago; he has been a wise counsellor and close friend all the years since. Others on Dr. Kiley's staff also have my gratitude for their assistance in seeing this project through: Dr. Joseph E. Goldberg, Colonel John C. Bordeaux, and Lieutenant Colonel Paul Taibl. Janis Hietala, my editor, deserves great credit for her editorial skills and her superb design work—I thank you.

The reader will kindly find me alone at fault for any failings in argument, any inaccuracies in facts, and any stylistic infelicities—read: Any errors are unfortunately my own. All others should be held blameless—except my Uncle Joe who says he taught me everything.

ROBERT C. HUGHES

SDI: A View from Europe

1. TO AVENGE OR TO DEFEND?

Although there now appear to be insurmountable difficulties in an active defense against future atomic projectiles similar to the German V–2 but armed with atomic explosives, this condition should only intensify our efforts to discover an effective means of defense.
—General of the Army H. H. Arnold
The War Reports

There is not now, nor will there ever be, a single Strategic Defense Initiative (SDI) system of comprehensive ballistic missile defenses—as envisioned by President Reagan—to protect the territory of the United States and that of its allies. Were it not for political controversy, policy chicanery, and extravagant claims that warped SDI from its outset, this conclusion would not have to be stated at the beginning of this examination of the implications of SDI for NATO's strategic concept and for Alliance security. That said, however, in all likelihood the United States will eventually deploy a limited system of active defenses against ballistic missiles. Given the continuing proliferation of intermediate and short-range ballistic missiles, the United States will have to develop such a system even if damage limitation were the only rationale. And it is not.

The SDI debate all too often proceeded on both sides of the Atlantic from remote starting points as well as fantastical assumptions: for example, that effective space-based missile defenses could be deployed in the early 1990s, using off-the-shelf technologies; that in the not too distant future, if the technologies pan out, a "thoroughly reliable" SDI system will protect America and its allies and render ballistic missiles (if not "nuclear weapons") "impotent and obsolete"; or that the whole idea of SDI is a harebrained, dangerous distraction of the Alliance away from the military forces needed to carry out a strategy of flexible response, away from opportunities for arms reductions through negotiations, and away from the nuclear forces, theater and central systems, needed to underpin deterrence and to couple Europe's security with America's.

POINT OF DEPARTURE

In contrast to speculative assumptions, this book takes as its starting point certain concepts: SDI projects might yield technologies that can be translated several decades from now into a series of layered systems with limited capability to defend the United States and its allies against ballistic missiles. It is highly unlikely, however, that a single system of comprehensive missile defenses—or even a system of systems—will ever evolve. The technologies, no doubt, will develop in evolutionary, incremental ways, not through revolutionary phases (though there will be some dramatic breakthroughs), with the applications to conventional defense capabilities just as important as those for strategic defenses. And, if a future administration decided to go ahead, deployments would occur over decades and would be folded into US modernization of nuclear and conventional forces, arms control reductions, and Soviet developments in offensive and defensive systems.

The deployments would also occur in the context of new military doctrines and strategies in NATO.

No matter how often the president or other senior administration officials in the United States and in Europe declared it, the point that the SDI was a research program never stuck for long in the public debate. The same officials themselves would undercut the issue with speculation about deployment and production schedules. Unfortunately, there was all too often a precipitous jump from the escarpment of research into wonderful but unsubstantiated claims, by supporters and critics alike, about what SDI might or might not contribute to crisis stability, to deterrence of war, to damage limitation, to war termination, and to nuclear arms reductions between the superpowers.

Some saw in SDI the collapse of the Alliance arrangements of extended deterrence, an isolationist yearning of the United States to protect its own borders and to reduce its risks, along with consequent increased potential for conventional war in Europe. In contrast, others welcomed the president's attempt to slip free of the nuclear dilemma that had threatened annihilation of civilization as well as his attempt to have the capability "to defend" rather than "to avenge." The "proper" response to the president's disjunctive question about choosing vengeance or defense must surely be, "I'd rather defend." But those were, of course, not the only alternatives of the choice, nor was the question the right one (except politically).

Paradigms of Deterrence. As Bernard Brodie recognized, the paradigm for US thinking about war changed forever with the explosion of the first atomic bomb: "... everything about the atomic bomb is overshadowed by the twin facts that it exists and that its destructive power is fantastically great," he said. The concepts of offense and defense when applied to deterrence became conflated and barely distinguishable from certain angles. As the nuclear age proceeded by the fits and starts of modernization and stockpiling, and as the Soviets reached parity in the late 1960s with the United States in nuclear arsenals, the ultimate guarantee of deterrence became not so much the threat of massive retaliation but the assurance of mutual destruction. This threat became increasingly less credible when extended to collective defense arrangements with Western Europe, beyond retaliation for attack on the US homeland.

In the search for assured security, with SDI President Reagan reached for a new paradigm, a paradigm born of science and arms control, as well as of frustration with modernization and with the threat of mutual assured destruction. No "new" weapons and no "new" arms reduction schemes, however clever, held any promise of slipping the bonds of the nuclear threat as the final guarantee of security. For Reagan, any "hope" for escape was better than simply the promise of more of the same forever. No more desks to crawl under, no more shelters to find in a firestorm, and no more shovels to pile up three feet of earth (vice six). In the jumble of human wishes and in the jangle of ugly nation-state realities lay the motivations for SDI. As the president asserted, the effort presented "a vision of the future which offers hope." It was this dimension of SDI that caught the imagination and caused most of the controversies for those who saw "hope" only in traditional, more-of-the-same approaches.

Chewing Over the Many Explanations of SDI.

Uncertainty about SDI's objectives, equivocation in terminology, and ambivalence about the consequences of strategic defenses plagued Reagan's initiative. Confusion swirled around the many explications of how SDI, the concept, not necessarily the technologies, can help the West escape nuclear dilemmas. Much has been written elsewhere on the origins of SDI and on the Reagan administration's attempt to protect the concept from shredding by US and Alliance bureaucracies.[1] The United States broke one of the NATO rules—no surprises—by not introducing the ideas at least in capitals well before announcing the initiative publicly. Most cabbage is "twice chewed" in Brussels and in capitals before it can be digested.

The name of the program itself caused contentious reactions. In March 1983, the president did not use the term "Strategic Defense Initiative." The president challenged scientists, arms control negotiators, and strategists "to break out of a future that relies solely on offensive retaliation" for security; he talked about the "human spirit" rising above dealing with other nations and human beings by threatening their existence. Relying on the specter of retaliation for deterrence, the president said, was "a sad commentary on the human condition." Some responded, "And so it must remain."

President Reagan invited the nation to "embark on a program to counter the awesome Soviet missile threat with measures that are defensive." Right after the president's speech, the press named the program

"Star Wars," a label the administration could never remove. By the fall of 1983, the name "Strategic Defense Initiative" began to gain currency in the government. The name was captured in the bureaucratic lexicon with James Fletcher's Defensive Technologies Study and the policy studies led by Franklin C. Miller and Fred S. Hoffman, as well as with the establishment of the SDI Organization (SDIO) in January 1984, chartered in April 1984 under Lieutenant General James A. Abrahamson, USAF, its first director.

The president recognized that "defensive systems have limitations and raise certain problems and ambiguities," and he acknowledged that "if paired with offensive systems, they [strategic defenses] can be viewed as fostering an aggressive policy." At the same time, he could not have been prepared for the persistent confusion surrounding what he hoped would be a straightforward effort, as he said, "to define a long-term research and development program to begin to achieve our ultimate goal of eliminating the threat posed by strategic nuclear missiles."

What's in a Name? The word "strategic" generated questions from the allies about whether the United States was in fact interested only in protecting its own homeland through a global, space-based system, despite what the president had said. After all, in an arms control context, the term "strategic" when applied to ballistic missiles referred to the SALT limits of 5,500 kilometers—roughly the range of the Soviet SS–5 missile. In effect, this range defined the difference between strategic and nonstrategic systems.

The concept behind the term *strategic* begged for clarity since the United States seemed to have only, or at least primarily, the SS–17s, SS–18s, and SS–19s in mind in the explanations of what warheads, launchers, and missiles—terms not carefully distinguished—SDI was supposed to defend against. The Europeans at first heard little from US officials about the SS–20s in the context of strategic defenses. The concept of strategic in US thinking was in contrast to *theater* or *tactical*. However, from Europe's viewpoint and—particularly in the arms control context—from the Soviet viewpoint any weapons that hit European or Soviet territory would be as strategic as those striking US territory.

Given a persistent myth in European defense thinking that somehow Soviet strategic systems are targeted solely against the United States and not against Europe and that given the new threat to Europe (before the Intermediate-range Nuclear Forces Treaty) from shorter range Soviet ballistic missiles such as the SCUD–Bs, SS–12/22s and SS–23s, strategic defenses did not appear to the Europeans to increase their security. In fact, SDI might indeed lead to instabilities in areas such as first strike incentives, the arms race, and crisis management. For these reasons, the then German defense minister, Manfred Woerner, sought and applauded the US discussion of possible applications of the SDI technologies against the SS–20s. Woerner did so on the margins of the Nuclear Planning Group (NPG) ministerial meetings held in Luxembourg on March 26–27 in 1985 and in Wurzburg, West Germany, on March 20–21 in 1986. At the Wurzburg meeting, NATO also expressed concern about Soviet deployment of the SS–23. Whether defenses could ever work appeared less important than the demonstration to European governments and publics that the United States had not abandoned its commitment to extend its defense and deterrent capabilities to Europe.[2]

Besides the difficulties with the word *strategic*, the Soviet negotiators in Geneva, as well as respected defense commentators in the West, suggested that strategic systems could include "defenses" that might be used to increase the capabilities of the "offense."[3] For example, with thoroughly reliable defenses against ballistic missiles in place, one or the other of the superpowers might have increased incentive to strike first, believing that its own intact defenses would be able to defend against or even to preclude a ragged retaliatory strike.

In other words, the Europeans feared that one superpower would try to attain superiority through strategic defenses. The Europeans did not want the Soviet Union to be that superpower, nor did they want the United States committed to that course. Nuclear parity (perhaps at lower force levels) in perpetuity was the best that Europe could hope for. Deterrence by forward defense and the ultimate threat of retaliatory nuclear punishment were working fine for Europe, and there was no sufficient reason to shift the paradigm to deterrence through denial and defense.[4] A "Program for Deterrence of Ballistic Missile Attack" that included defenses against ballistic missiles of all ranges might have been better received conceptually in Europe than was the Strategic Defense Initiative. That is not a suggestion, however, that the whole issue was merely a matter of semantics; it most definitely was not. SDI struck the issues at the core of the transatlantic bargain.

As the debate related to the terms *strategic* and *defense* continued, there was also difficulty with the idea of *initiative*. The United States was slow in pointing to the Soviet build-up in strategic offense and in strategic defense as the main reason for the program. By early 1985, however, the United States was doing a better job.[5] From a NATO viewpoint, the concept should rightly have been that of a Western response to Soviet activities. In fact, there was more discussion in Europe about how the Soviets would "react" to SDI than about how the West should respond to the prior extensive and advanced Soviet efforts. If the United States had presented SDI as a response or even as a needed acceleration of limited research under way, the reception for SDI in Europe might have been cold but not icy.

OBJECTIVES OF SDI

The label on the program was not the only contributor to the confusion. The very notion of reopening the strategic defense debate was, of course, itself highly controversial, especially since the allies thought the matter had closed with the Anti-Ballistic Missile (ABM) Treaty of 1972.[6]

What was clear from the analyses and recommendations of the study efforts led by Fred S. Hoffman (in contrast to that led by James A. Fletcher), notwithstanding protestations about layered defenses, was that SDI would in the first instance investigate the technologies for defending high-priority military assets, primarily through the terminal and late mid-course parts of the trajectory. In other words, in early stages of the initiative, strategic defenses could be added to the forces that underpin deterrence as the West has known it. This concept was quite different from the challenge of the president to find ways to defend people from ballistic missile strikes. The defense bureaucracy had, in effect, already determined the answer to the president's challenge before the research was even under way.[7]

Some Europeans speculated that the real US purpose was regaining nuclear superiority through the edge the West enjoyed in certain technologies. Apparently concerned about US intentions, Prime Minister Thatcher, on December 22, 1984, reached an understanding with President Reagan on four points: superiority was not the purpose of SDI; deployment would be a matter of negotiation with the Soviets; the aim of SDI was to enhance deterrence; and negotiations with the East should aim to achieve security with reduced levels of offensive systems on both sides.

What, for that matter, was the problem SDI sought to correct? If SDI was meant to correct the vulnerability of US land-based ICBMs to massive attack by highly accurate Soviet land-based ICBMs, then were all the other measures for increased survivability insufficient? Even, for example, the small mobile missile that the Scowcroft Commission had recommended in its 1983 report? Or the deployment of the MX with a mobile capability, perhaps aboard railroad cars? Moreover, there were also under way the development and acquisition of the Trident submarine, D–5 warheads, deployment of 50 or more MX, and cruise missiles—all these would add significantly to the assured survival and retaliatory capability of the US nuclear arsenal. What then was the main reason for SDI?

The Europeans rightly understood that there was much more to the motives for SDI than improvement in the survivability of the land-based missiles.[8] In short, through SDI and other means, the United States intended to shift its own deterrence strategy and eventually that of the Alliance toward more reliance on systems that were defensive, non-nuclear, standoff, and discriminate (whether nuclear or conventional).[9] In addition, there was the desire to devalue to nothingness (if possible) the worth of ballistic missiles; these were after all the most dangerous of the Soviet nuclear weapons.[10] These weapons meant that the American president would have too little time to react in the event of attack; launch on warning was not a desirable defense posture—least of all in public.

Moreover, ballistic missiles with conventional warheads would be in the hands of a large number of nations by the early part of the next century. Something would have to be done about defending against such missiles, no matter what the allies might think about the inviolability of the ABM Treaty and about BMD research. By the spring of 1988, for example, the city-to-city ballistic missile attacks (numbering in the scores) in the Iran-Iraq war, as well as the Saudi purchase of Chinese ballistic missiles capable of reaching the major cities of the Middle East, were just more examples of proliferation of the missiles since SDI began and of the role such weapons might play in conflicts.

The Reagan administration was consistent in this pattern of thought about devaluation or even elimination of ballistic missiles and about discriminate conventional weapons: from the INF zero-zero proposals, through the president's SDI speech in 1983, to the Follow-on Forces Attack (FOFA) operational concept, to the SDI program, and finally to the Reykjavik summit.[11] At that summit, President Reagan and General Secretary Gorbachev spoke of the elimination of ballistic missiles within ten years, much to the horror of the Europeans, especially the French and the British. As then Assistant Secretary of Defense for International Security Policy Richard Perle argued in Europe, the primary threat to the United States from the Soviet Union was the massive Soviet land-based ballistic missile nuclear arsenal. To the extent that the arsenal could be reduced quantitatively or devalued through arms control, increased survivability of US offensive systems, and development of strategic defenses—to that degree—stability would be fortified.

Europeans were concerned, however, that the United States, in its haste to devalue ballistic missiles, might play into the hands of anti-nuclear political forces and contribute to the "de-nuclearization" of Europe. This concern was especially acute when senior spokesmen in the United States, including the president, began to speak of the immorality of deterrence based on the threat of mutual assured destruction. This theme contributed, for different reasons, to the anti-nuclear sentiments in all parts of the Alliance.

In addition to confusion about SDI terminology, objectives, and intentions, there was a palpable concern that no matter what objections European governments might raise, the United States had already decided to deploy space-based strategic defenses against ballistic missiles. The US explications justifying SDI, at least in the eyes of some European experts, continued to shift, to bolster and validate the prior decision to go ahead. Given this context, European political leaders were probably not so surprised as they pretended when senior officials of the administration in the fall of 1985 and thereafter began to talk about early deployment of strategic defenses. However suspicious and forewarned by their own insights they may have been, Europe's political leaders reacted strongly and demanded that the United States keep its word in the Reagan-Thatcher points of December 1984—namely, that SDI-related deployment would in view of the treaty obligations have to be a matter for negotiations with the Soviet Union and consultation with the allies. The question of deployment was closely linked to contentious issues about the so-called "narrow" and "broad" interpretations of the ABM Treaty.[12]

The protests of the European allies were eventually quieted with the assurances by Secretaries Weinberger and Shultz that although the broad interpretation was the "legally correct" one, it would not be necessary to restructure the SDI experiments (designed to comply with the narrow interpretation) during the remainder of the Reagan administration.[13] That agreement did not, however, satisfy Senator Nunn, who launched his own investigation into the legal and political meaning of the treaty, including Agreed Statement D, as well as into the constitutional significance of Senate advice and consent to treaty ratification and to subsequent interpretation of individual treaties. The Nunn study included the negotiating and ratification records, as well as a review of the text of the treaty and the record of compliance after the treaty was signed in 1972. The study upheld, of course, the traditional or narrow interpretation of the ABM Treaty.

NITZE CRITERIA

Along with the administration's pledge not to deploy strategic defenses without prior consultation with the allies and negotiation with the Soviet Union, as well as its pledge not to proceed with the SDI tests under the "broad" interpretation of the ABM Treaty, the allies also clung to the so-called Nitze criteria for deployment. As Ambassador Paul Nitze stated repeatedly in the spring of 1985 and thereafter, the president had directed that the object of SDI be "to provide the basis for an informed decision, sometime in the next decade, as to the feasibility of providing for a defense of the United States and our allies against ballistic missile attack."

In the context of Soviet efforts on offensive and defensive strategic systems, Nitze identified two "demanding" criteria to judge the concept of SDI feasibility. Such defenses must be "reasonably survivable" or they would be tempting targets for a first strike. Second, the new defensive system must also be "cost-effective at the margin—that is, it must be cheaper to add additional defensive capability than it is for the other side to add the offensive capability necessary to overcome the defense."[14] Some pro-SDI commentators suggested that these criteria were a way to ensure that there never would be any deployment of SDI-derived strategic defenses.

With the Nitze criteria well-established by mid-1985, Secretary General Carrington and other leaders of the Alliance could rest more confident regarding what Carrington called the "firebreak" between research and development of technologies on the one hand and the development and deployment of prototypes and components for strategic ballistic missile defenses on the other. Although the word *research* never appears in the text of the ABM Treaty, the term was often used loosely to describe activities short of prototype development, testing, and deployment. Carrington feared that any US unilateral moves toward deployment of strategic defenses would cause grave problems within the Alliance, since there was no consensus among the nations on anything beyond the investigation into SDI technologies.

SDI: CONTRARIETIES

The SDI debate in Europe and in the United States has been replete with confusing quirks of thought and expression. Small wonder that the public might be confused when defense "experts" seek to clarify the place of strategic defenses in deterrent theory but only add to the difficulty with contradictory terminology and logical contortions. The president took to "speechifying" when he asked these supposedly simple questions: "Wouldn't it be better to save lives than to avenge them?" and "What if free people could live secure in the knowledge ... that we could intercept and destroy strategic ballistic missiles before they reached our own soil and that of our allies?" and "Are we not capable ... of achieving a truly lasting stability?" Despite cheers from the right-thinking crowd proud to answer boldly for the president, these questions do not yield straightforward responses. Simple answers may offer escapes from painful dilemmas, but not without putative costs in clarity and in comprehensiveness.

Among the many quirks to be confronted when assessing SDI's implications is the assumption that *offense* refers to deterrence while *defense* refers to war proper or to the modern term, *warfighting*. In a way this argument is a variation on the discussion between Secretary of Defense McNamara and Prime Minister Kosygin in Glassboro, New Jersey, in 1967.[15] In those talks, McNamara argued that the proper response to the Soviet deployment of strategic defenses along with their modernization of offensive forces would be for the United States to expand its "nuclear offensive forces." At that, Kosygin "absolutely erupted. He became red in the face. He pounded the table. He said, 'Defence is moral, offence is immoral.'"[16]

In a mid-1985 BBC interview, McNamara asserted that Kosygin's counterparts today would take what was the US position then. In an oxymoronic twist, the Reagan administration itself argued what was at Glassboro the Soviet position, including the suggestion that deterrence based on the offensive threat of nuclear use is immoral. This shift of positions occurred in less than two decades. It is no surprise, therefore, that Robert McNamara and the other members of the "Gang of Four" (Gerard Smith, George Kennan, and McGeorge Bundy) have striven mightily against SDI from its outset. They have had long-term interest in assured destruction (based as it is on offensive forces and territorial vulnerability, a concept that should exclude or at least give short shrift to defenses).

The Soviets and nearly as often the allies also expressed strained argument that SDI defenses would not work and therefore should not be pursued and in the same breath sometimes argued that SDI might work partially and therefore should never be deployed. If deployed by the United States, presumably strategic defenses might upset the stability of deterrence and cause the Soviets to escalate the arms race and to deploy their own defenses.

Senior Reagan administration officials might be forgiven for not demonstrating sufficient gratitude for the concern shown by the allies (and even the Soviets on occasion) that the United States might be wasting its money in pursuing SDI technologies. Neither the Soviets nor the allies have been so solicitous of the US Treasury in the past. Even the most credulous Atlanticists might wonder whether there might not be more to the objections.

While downplaying any military merit in SDI, the Soviets along with the NATO allies were nonetheless worried about the potential leaps the United States might take in commercial civilian applications of technologies in areas such as high-speed computers, materials, miniaturization, command and control, and lasers. In addition to the economic potential in SDI of keeping the United States at the cutting edge of technology, the Soviets appeared most concerned because SDI struck at the heart of their status as a superpower, namely their land-based ICBMs, the main part of the Soviet nuclear arsenal. If indeed the United States were to devalue the significance of ballistic missiles, the worth of the Soviet Union as a superpower would also thereby be diminished.[17] As Benjamin Lambeth and Kevin Lewis stated in a 1988 article, "The Kremlin and

SDI," by far the "most likely source of Soviet agitation over SDI has to do with high-level concerns that continued progress of the U.S. program may undermine worldwide appreciation of Soviet military prowess, irrespective of any technical problems SDI may encounter along the way." They also suggested that Soviet boasts of easy countermeasures have had "a tone of nervous whistling in the dark."[18]

Another of the seemingly contrary complaints about SDI was how dangerous strategic defenses would be if they were imperfect and how dangerous they would be if they were "perfect" or "thoroughly reliable"—a phrase several senior administration spokesmen used frequently in 1985 to describe the operative objective of the research. The US spokesmen did not claim that strategic defenses by themselves would need to be "perfect" against every and any threat from ballistic missile attacks against the United States and its allies. Instead, as the argument went, SDI needed only to create "sufficient uncertainty in the mind of a potential aggressor concerning his ability to succeed in the purposes of his attack" that he would be deterred.[19] In short, SDI was meant to enhance and not to replace deterrence. However, President Reagan also made it clear repeatedly that "we must seek another means of deterring war. It is both militarily and morally necessary."[20]

Once the allies recognized the degree of presidential, congressional, and popular support for SDI in the United States, they successfully worked to fence off SDI research proper from the deployment issue and gave minimal assent to the research alone. In other words, Prime Minister Thatcher, Chancellor Kohl, President Mitterrand, and (a bit later)

Prime Minister Chirac endorsed the research as a hedge but not deployment of strategic defenses. The Alliance's instruction to the United States was, in effect, "Go ahead and do the research which our industries will help you with as long as it does not cost us anything and as long as we get the benefits of the technologies we work on. By the way, remember we have not agreed to deploy any systems."

There is not a little irony in the argument of those who favor deployment of strategic defenses primarily to protect military assets. For example, if such defenses could successfully protect retaliatory strategic forces in missile fields, in nuclear submarine pens to some extent, and at airfields for aircraft capable of delivering nuclear weapons, the potential enemy might decide to shift his targeting objectives in order to hold "soft targets" like unprotected cities even more at risk. Although Robert S. McNamara coaxed both superpowers and the allies from the concept of massive nuclear retaliation to assured destruction and flexible response, the new defense situation could again bring to the fore a potential increase in strikes against cities, a potential that will always be there as long as there are weapons of mass destruction.

Survivability versus Vulnerability. One of the prickly issues surrounding SDI concerned "vulnerability" versus "survivability" of strategic forces, a dilemma complicating decisions on strategic modernization, negotiating positions in arms control, and the efficacy of any strategy relying on retaliation. As called for in the McNamara approach, deterrence depended on sufficient, survivable offensive nuclear forces to retaliate, primarily against leadership, communications, and military targets, in the event of a nuclear strike by the enemy.

In this equation, defenses to protect the population and even to defend nuclear forces played an ever diminishing role. The McNamara-Johnson thesis of mutual assured destruction (MAD) found expression during the first Nixon administration in the ABM Treaty of 1972. From the mid-1970s on, missile defenses were thought not to contribute to first-strike stability. If the Soviets wished to go ahead, and they did, of course, with their own extensive air defenses, civil defenses, and even limited ballistic missile defenses around Moscow, such defenses according to the McNamara deterrent theory would not change the balance. The situation remained stable as long as US and Alliance nuclear forces could still penetrate in sufficient numbers to attack valuable targets in the Soviet Union.

Ambassador Paul Nitze rejected McNamara's belief that the ABM Treaty codified MAD doctrine. In a mid-1985 British Broadcasting Corporation (BBC) interview, Nitze and interviewer Michael Charlton had the following exchange:

> NITZE: There were no common understandings [with the Soviet Union in the SALT 1 and ABM treaties about "basic issues"]. We agreed on the language of several specific documents.
> CHARLTON: So the idea that the ABM Treaty reflected a common doctrine ...
> NITZE: ... is nonsense.
> CHARLTON: ... among the two powers ...
> NITZE: ... is nonsense. It's nonsense.
> CHARLTON: ... of "mutual assured destruction" is not what you accept?
> NITZE: It is nonsense, nonsense.

Nitze did acknowledge that at the time of the ABM Treaty ratification, there was "a good deal of talk" about whether there was agreement to the "spirit of something that went beyond the pieces of paper—we hoped that they [the Soviets] would look at it that way. They did not."[21] The Soviets, he asserted, made it clear that they were agreeing to nothing beyond the specific obligations of the treaty language.

The survivability and vulnerability issues at the core of deterrent and defense dilemmas of the nuclear age will not be resolved; they can only be managed. The thrust of the ABM Treaty prohibiting territorial defense against strategic ballistic missile attack meant that vulnerability to attack and survivability of retaliatory forces would remain key conditions of deterrence. What is often not clear in discussion of these concepts is that while the United States should continue to ensure its own survivable forces and to work for a vulnerable Soviet Union, there is no commitment to keeping the United States vulnerable forever.

As Ambassador Nitze stressed, the Soviets never gave up on defenses. Moreover, the reason the United States did so itself was that the technologies did not hold sufficient promise and were too expensive. In other words, the United States agreed to the ABM Treaty not because of some compelling paradigm or some rationalist theory of Herman Kahn's and Robert McNamara's, but because the United States wanted to ensure the effectiveness of its retaliatory nuclear forces against Soviet targets. To do so, the United States had to limit Soviet ballistic missile defenses in quantitative and locative terms.

What SDI Is Not. From January 1985, when Ambassador David M. Abshire (then permanent representative of the United States to NATO) was named the administration's "point man" in Europe on SDI, through to the Moscow summit of mid-1988, US officials spent as much time defining what SDI was not as they did defining what SDI was. A cottage industry of issues for defense "experts" to grapple with attended the establishment of the research and development efforts labelled "SDI." Each expert offered his or her own insight into what SDI really was and into SDI's possible implications for everything from conventional warfare and nuclear deterrence, through the price per pound of space lift and the cost of computer chips, to the economic dominance of Europe by the United States and the relegation of the European armaments industries to perpetual serfdom.

Anything near a complete description of what US spokesmen had to deny as the often fanciful, yet said to be "real," objectives, intentions, timetables, assumptions, and rationale for SDI would be tediously long and of little value. A few examples suffice to provide some inkling into the scope of the many misconceptions. One conceptual error that afflicted supporters and critics alike was the reification of SDI; that is, SDI too often was discussed as though it were a single system either already in existence or soon to exist. SDI is not a single system but a series of more or less interrelated projects of research into a variety of technologies, with some of the projects much older than SDI itself. What the SDI Organization (SDIO) added to the technology investigations was a focus on the objective of defending against ballistic missiles, integrative and innovative management, and with congressional support, large increases in funding.

Besides discussing SDI as an "it," as though "it" were a separate weapons system like a tank, the public debate tended to isolate SDI from the context of NATO's modernization of military forces, arms control negotiations, and Soviet efforts to improve their offensive and defensive strategic forces. In isolation, SDI might well have looked like an extravagant US excursion into areas of research that could promise, if pursued, nothing except instability. When discussed apart from modernization of strategic nuclear forces, attempts to reduce strategic arsenals by 50 percent, and Soviet efforts in strategic defenses, SDI loses its explicit reason for being—to decrease the vulnerability of the United States to Soviet nuclear attack by devaluing the worth of ballistic missiles. Too often SDI was discussed as though SDI were a panacea for all strategic questions, rather than one alternative to respond to one issue in the deterrent equation.

With a finger pointed at Soviet efforts, the United States kept the political heat on the governments of several NATO nations so that they would not break the consensus on the importance of the SDI research as a prudent hedge against a Soviet breakout from the ABM Treaty. Although nations like the United Kingdom agreed, they were also careful not to allow the United States to make too much of clear violations of the ABM Treaty such as the large radar at Krasnoyarsk. The British, for example, worried that the United States would prematurely build the case for deployment of strategic defenses to offset a potential Soviet territorial defense.

The Reagan administration did not need the approval of the allies, only their acquiescence, in order to maintain the support of the US public for SDI. After all, the public would not think kindly of the European argument that the Alliance would be better off if the United States remained vulnerable to annihilation through Soviet nuclear attack, especially attack with ballistic missiles. This thinking was abhorrent to conservative national security analysts like Jeane Kirkpatrick, who saw vulnerability growing in significance: "The vulnerability of the United States [to ballistic missiles] is *the* most important fact of our times. Most Americans still do not understand that improvements in the accuracy and speed of Soviet missiles and the silencing of Soviet submarines have rendered the United States more vulnerable than at any time in its history."[22]

Deployment of Ballistic Missile Defenses. Yet another cause for confusion and equivocation in the explications of SDI concerned deployment. The administration's early promises that President Reagan would make no deployment decision in his term of office were copied onto vellum scrolls and carried from security conference to security conference throughout Europe from early 1985 on.

To provide a context for SDI, in January 1985, a White House pamphlet on SDI included discussion of the relationship between modernization and SDI research: "In the event a decision to deploy a defensive system were made by a future President, having a modern and capable retaliatory deterrent force would be essential to the preservation of a stable environment while the shift is made to a different and enhanced basis for deterrence." Such an environment would also include arms control efforts to reduce nuclear arsenals during any transition to deterrence based more and more on defensive systems.

Even given this assumption that SDI deployment was some time away, by the fall of 1985 the administration was nonetheless promoting the "broad" interpretation of the ABM Treaty. By late 1985, public remarks by senior officials left little doubt among the allies that the United States would decide quickly, perhaps unilaterally if necessary, to deploy strategic defenses early. Such deployment would certainly occur even if there were promising technologies only for terminal defenses of military assets, a focus that was not SDI's principal thrust.

MULTIPLE PERSPECTIVES OF THE
UN-DEFINITION OF SDI

The Reagan administration sought not to get hoisted with its own petard in discussions of what SDI was and what it was not. At the same time, there were so many "spokesmen" in and out of government presenting contradictory versions of SDI that the allies rightly did not know which version to deal with.

In the fall of 1986, Charles Krauthammer, in a *New Republic* article, pointed up this difficulty when he identified four SDIs. First, the president's vision of strategic ballistic missile defenses, based in space, that protected populations from attack and changed the nature of deterrence. Then, there was the bureaucratic version that turned the president's "vision" into a program to develop and deploy business-as-usual, land-based terminal defenses of

US land-based ballistic missiles and possibly other strategic assets. Third, in connection with the Reykjavik summit, there was the "insurance" version of SDI that President Reagan said would be deployed after the elimination of ballistic missiles in the arsenals of the superpowers. Last was the "Soviet nightmare of post-ballistic, space-based, high-tech *offensive* weapons, like particle beams or rail guns, which might be developed by the United States under the pretext of developing defensive weapons." As Krauthammer remarks, this last version was disingenuous, since such weapons could be developed and deployed without any pretexts and indeed not fall under any ABM Treaty limitations.[23]

Consequences for NATO. The Strategic Defense Initiative has as much to do with lessening the political value of nuclear weapons as it does with a military capability to defend against ballistic missiles—whether such missiles be targeted against population centers or military assets or both. From the late 1950s, Europeans have worried that the United States would tire of its extended deterrence obligations and look for ways to decouple itself from the security of Europe. General de Gaulle in the early 1960s rejected—or perhaps just "pocketed"—the assumptions of extended deterrence and hence NATO's strategy of flexible response. The SDI was now one more pressure for the Americans to decouple from a European commitment, like the threat of troop withdrawals and burdensharing complaints.

The fragility of NATO's strategy of flexible response had been evident to defense experts and political leaders for a long time—Henry Kissinger in the 1970s and Robert S. McNamara in the 1960s.

The strategy issue became so sensitive a topic at NATO headquarters, especially after Kissinger's critical speech in Brussels in 1979 on the margins of NATO's celebration of its thirtieth birthday, that any suggestion for a review of that strategy immediately met with strong opposition. The delivered text of Kissinger's speech was even more strongly worded than the sanitized public version reported on since 1979. In 1984, NATO reaffirmed the tenets of East-West relations laid out in the Harmel Report of 1967 and thereafter would not indulge itself in any reevaluation, even to commemorate the fortieth anniversary of the Alliance, lest the "wrong" answers might evolve. When the United States decided to deny calls from some nations for a formal NATO analysis of SDI, for example, it found several backers among the allies; even a hint that the strategy itself might be examined in such a study was enough to ensure inaction.

The implicit understanding was that if senior US officials would stop stating that SDI would change NATO's strategy and that the current deterrent strategy was immoral, then the Alliance consensus would hold on support of SDI research. The president's program struck at the core of deterrence. The rationale that SDI would enhance deterrence, one of the points President Reagan agreed with Prime Minister Thatcher in December 1984, was helpful to consensus. However, any suggestion that US vulnerability would be lessened was not.

The allies wanted to see the evidence that through SDI the security of all NATO nations would improve, not just that of the United States and Canada. While nuclear weapons had rendered war in Europe unthinkable, the fear now was that war, conventional or nuclear, might be thinkable if the two superpowers were to become less vulnerable to attack from one another. It did not seem to matter that even if SDI were perfectly effective, there would still be thousands of warheads that could be delivered against the homelands through a number of delivery modes. Simply put, SDI would not mean a Europe or a world without nuclear weapons.

Transitions. The questions Alliance nations wished answered concerned the consequences for strategy, stability, and security if the United States alone were to deploy strategic ballistic missile defenses, if the Soviets alone did so, if both sides did but at different paces, and if both sides did in a managed transition. This issue of transition—the management of strategic defense deployments, the modernization of offensive nuclear forces, and deep reductions in the nuclear arsenals through arms control—loomed large in the list of concerns in thinking through the future of NATO's strategy.

The thought that the president of the United States dared to imagine a time when there would be no nuclear weapons chilled European leaders who had allowed Alliance security to depend too heavily on nuclear weapons from the mid-1950s to the present. The worst European nightmares became reality when President Reagan and General Secretary Gorbachev discussed such a world at the Reykjavik summit in the fall of 1986: the elimination of all ballistic missiles in ten years, as well as a long-term commitment to a world without nuclear weapons.

There have over the years been introspective debates in Brussels about the efficacy of NATO's strategy. However, as Lawrence Freedman remarked, "The emperor deterrence may have no clothes, but he is still emperor."[24] The deterrent concept has remained questionable despite Alliance attempts to make flexible response more credible through measures such as acquisition of discriminate and accurate weapons, as well as through declaratory policy focus on military, leadership, and economic targets—for example, the refinement of Secretary James Schlesinger's work on the Single Integrated Operational Plan (SIOP) through President Carter's directive, PD 59, and President Reagan's national security decision directive, NSDD 13.[25]

In sum, the allies raised questions about SDI that were not answered in a thoughtful and systematic way. The NATO members may continue to avoid formal reexamination of Alliance strategy, even in light of the changing politico-military context in Europe. However, whether or not there is any formal reassessment, there remains a need to examine potential implications of strategic defenses. European nations will need a roadmap—or a conceptual framework, to use a phrase with currency—that shows how such defenses make the way ahead easier, more efficient, and more stable than the present deterrent arrangements based on the threat of the use of nuclear weapons.

Although political elites only reluctantly acknowledge it, in many Alliance nations support continues to erode for nuclear deterrence, with opposition very strong in a few of them. At the same time, there is anxiety that America may be tinkering with NATO's strategic concept before there is any certainty that the combination of strategic arms reductions, the modernization of offensive systems, and deployed defenses would yield a better combination of forces than the present forces to underpin deterrence.

SDI: The Vision. At one and the same time, the administration touted SDI as the beginning of a revolutionary strategy of defense by denial and the enhancement of the old strategy of flexible response. This confusion helps explain the multiple rationales for SDI.

When the president asked scientists, strategists, and arms control negotiators whether there might be a better way for deterrence to work, some answered immediately that there was no better way. However, many others took a more cautious approach to await the evidence from research into promising technologies and analyses of SDI's implications for strategy and force structure. Many of the scientists and strategists who supported the president but not his "vision" took the position that they shared his views for the long term, but in the meantime the superpowers would have to live for decades with the status quo. What response European nations gave to the president's vision, collectively and individually, and what implications SDI might have for NATO strategy remained factors that would help determine the character of any transition to a new deterrent strategy based on defensive systems.

2. EUROPEAN CANDLING OF SDI

While there have been common trends in Allied reactions, particularly among political and military elites, the absence of a single voice irked those in Europe and Canada who saw in SDI an opportunity to draw together a distinctly "European" security consensus, as opposed to ragged reactions.[1] The worst reaction they feared has been close to what happened: namely, discrete responses over time, but never quite on time and never quite complete. Without doubt, discussing these European reactions risks oversimplification; however, an understanding of the major responses helps any appreciation of the more nuanced views of individual nations.

CRITICISM FROM EUROPE

The allies expressed a variety of concerns about SDI, and the United States has found several of these issues particularly difficult to resolve or at least to manage. Among these charges are the following: The United States continues to show its disdain for the integrity of Europe's voice in matters related to its own security interests by neglecting to consult properly before launching major programs and strategy proposals affecting the security of Europe[2]; President Reagan, for all the loose talk about the use of nuclear weapons in the early years of his administration, joined the pacifists in criticizing the current strategy and in supporting the goal of eliminating nuclear weapons;[3] and, finally, US moralism, belief in a technological escape from the nuclear stalemate, and the constant and unnecessary tinkering the Americans do—all of these together—have brought into further question the credibility of "mutual assured destruction."

The timing of the initiative was particularly poor from Europe's viewpoint. Various strains of US thinking about strategic defenses had coalesced at a time when Europe was preoccupied with the deployment of intermediate-range nuclear forces. From this vantage point, the allies looked at strategic defenses as a possible risk for potential arms control agreements and for Europe's own security. Strategic defenses against ballistic missiles had not been a significant part of this equation since the early 1970s, and Europe preferred it that way.

Alliance authorities and the United States had not rigorously addressed the implications of ballistic missile defenses for several decades. The NATO nations were reluctant to undertake an analysis for which direction and conclusions were not known in advance. Just as nations do not like initiatives sprung on them, nations also do not like surprises from studies, especially when the issues relate to national survival.

A number of questions were on the minds of the Europeans: What is the problem that SDI seeks to correct? Is deployment of strategic defenses the best way or the only way to correct the problem? What is the rush in moving the research along so fast? What is in it for Europe? *Cui bono?* What will the Soviets think and do in the arms control process and in reaction to the US program? Even if the technologies proved feasible, cost-effective, and survivable, the

Europeans still wanted to know whether in the end the Alliance would be any better off than at present with deterrence based primarily on the threat of offensive nuclear forces.

PHASES OF REACTION

The forerunners of current debates on the worth of defenses are to be found in the deliberations prior to ratification of the 1972 ABM Treaty. Those debates closely match today's. The difference, however, is that the United States was then arguing the case to its allies and to the Soviet Union against deployed strategic defenses and was threatening a build-up of offensive forces, should the Soviet Union press ahead with building a defense of its homeland.[4]

To the frustration of President Johnson and Secretary McNamara, Prime Minister Kosygin would have none of the US theories about "deterrence" and "vulnerability" during their meeting at Glassboro, New Jersey, in 1967. As McNamara relates it, "Kosygin could no more understand our reactions than we could understand his. The two sides had totally different views of the nuclear world they lived in at that time."[5] This same perception was shared by President Nixon; he believed "... the Soviet Union did not separate deterrence and defense, but oriented their planning towards their ability to fight and survive and win a nuclear war."[6]

One of the nuclear verities that the Europeans (as well as many in the United States) thought they could hang onto was the US and Soviet agreement in the ABM Treaty of 1972 not to defend their populations or their military forces beyond the one ABM site agreed to in the 1974 protocol to the treaty. However, less than a decade later, a US president was asking the question, "Wouldn't it be better to save lives than to avenge them?"—in other words, to deter and to defend by denial rather than to deter by threat of retaliation. For many of the European allies, the answer to President Reagan's rhetorical question, at least among political elite, was a resounding "No"—a rhetorical answer because everyone presumably would want to save lives if that were the real choice. Whatever else can be said about Alliance reactions, most responses have had both a predictably ethnocentric focus, with Europe the hub, and at best a regional rather than global perspective.

The NATO allies have gone through at least four phases of reaction: they went from surprise and shock at an uncoordinated assault on the status quo, through guarded political acceptance of SDI as a prudent research program and technological hedge against Soviet efforts, to consternation about US talk of possible early deployment and of a new "broad" interpretation of the ABM Treaty; finally the allies came to the belief that SDI would disappear through arms control, through the change to less enthusiastic US administrations, and through the crush of budget deficits in the United States. The Soviet Union appeared also to have reached that conclusion in 1989.

Although European reactions to the research have generally been neutral, proposals for the deployment of defenses, with the necessary modification or abrogation of the ABM Treaty, have met stiff opposition in Europe. It is also true, however, that a number of defense experts, retired military officers, and industrialists strongly support SDI research, as well as the need for strategic defenses. [7]

Phase 1—March 1983 to April 1984. In the March 23rd, 1983, speech proposing research into strategic defenses, President Reagan shared with his audience, in his words, "a vision of the future that offers hope"—namely, a future in which "security did not rest upon the threat of instant U.S. retaliation to deter a Soviet attack." While issuing the challenge of this "formidable, technical task" to "the scientific community," the president acknowledged that it would probably take "decades of effort on many fronts" and that in the meantime, "we must remain constant in preserving the nuclear deterrent and maintaining a solid capability for flexible response."

The president's words on SDI were carefully laced with conditions and cautions. President Reagan was also mindful of Alliance members in suggesting that defenses would protect "our own soil and that of our allies" and in "recognizing the need for closer consultation with our allies."

The initial reaction in Europe was considerable pique, if not outrage, that the United States would have launched such an effort without extensive examination of the proposals by military and political authorities in Alliance capitals and at NATO itself. Although there was some prior notification in certain capitals before the speech, nonetheless, consultation was minimal and perfunctory—an impression the then Supreme Allied Commander, Europe, General Rogers, expressed publicly. The same, of course, might be said for coordination with the bureaucracies of the State and Defense Departments, although the president twice mentioned the support of the Joint Chiefs for the effort.

The Europeans never quite forgave the affront—pretended or not. Over the years since, the United States was at great pains to consult bilaterally and multilaterally with the Alliance nations over SDI. Despite considerable efforts in consultations of every sort and at all levels (from government to military to industrial), the European bill of particular grievances unfortunately grew through "flaps" over the ultimatum cum invitation (although never intended as such by the United States) for Allied participation in the SDI program, the issue of the "broad" versus "narrow" interpretation of the ABM Treaty, the matter of possible early deployment of some layers of defense, and especially discussions at Reykjavik where the superpowers talked of European security without the Europeans.

Although a case could be made that criticism of US consultations was unfair, nonetheless, the perception of inadequate, if not pedestrian, exchanges on the central strategy issues remained an irritation. Events as they unfolded were blotting the otherwise good record the United States built up from early 1985 on: frequent consultations in Allied capitals, at regional conferences, and at NATO headquarters by the US negotiators in Geneva, by the secretaries and assistant secretaries of state and defense, and by Lieutenant General Abrahamson (the director) and others from the SDI Organization (SDIO) and the scientific community.

Through most of 1983 and early 1984 the Europeans were consumed with matters related to the deployment of intermediate-range nuclear forces

(INF) and with arms control talks then under way. The Soviets walked out of the talks in December of 1983. Not much note was taken, except by some European defense experts and a few journalists, about the completion in October 1983 of important studies on SDI that had been commissioned by the White House after the president's speech. The first of these studies was the "Future Security Strategy Study," with Fred S. Hoffman as study director. The second was the "Defensive Technologies Study," under the direction of Dr. James C. Fletcher.[8]

Although both studies supported pursuit of the initiative and highlighted technologies that might pay off, at the same time the bureaucratic interpretations of the studies turned the president's vision into something different from what he had had in mind. For example, in his cover letter to the unclassified summary of the Fletcher report, the then under secretary of defense for research and engineering, Richard DeLauer, asserted, "This Strategic Defense Initiative will provide future Presidents with an option to enhance our deterrence capability by basing it on a mix of offensive and defensive forces."[9] In DeLauer's own words, the goal of SDI had already shifted from the original vision. Europeans would have less to worry about with a technology initiative that would "enhance" deterrence and end up with a mix of offensive and defensive forces at some future time. The Fletcher report concluded that the "technological challenges of a strategic defense initiative are great but not insurmountable."[10]

However, the Hoffman study gave no comfort to the allies. The study concluded that "effective U.S. defensive systems can play an essential role in reducing reliance on threats of massive destruction that are increasingly hollow and morally unacceptable." [11] Had the United States given the Hoffman study any visibility in Europe, the allies might have choked publicly (as some did privately). Clearly NATO's strategy itself was under attack because of its increasing lack of credibility and because of its moral unacceptability. These ideas were heretical to the Europeans—especially with deployment of intermediate-range nuclear missiles under way.

In January 1984, the SDI Organization (SDIO) was established—with General Abrahamson named as its head in mid-April. The Fletcher and Hoffman studies, along with directives (e.g., NSDD–119, January 6, 1984) resulting from technology and strategy work done that winter and spring, provided guidance to the SDIO. The early efforts turned into fabled battles waged among the Washington bureaucracies. It proved to be a series of herculean challenges to identify and cull out relevant programs already under way in the Services and in national laboratories, as well as to establish an organization in the Office of the Secretary of Defense (OSD), separate from the Services that wished to manage SDI (and its funds).

Phase 2—May 1984 to Fall 1985. The period between May 1984 and fall 1985 witnessed both successes and failures for SDI in Europe. On the positive side, European reactions changed from initial irascibility about the lack of consultation to a wait-and-see attitude. Perhaps the Americans would let the initiative die of its own financial and strategic weight. However, there was a gradual realization that senior administration officials, even if not the bureaucracies, were serious in efforts to focus on and to commit money and talent to the problems of defending against strategic ballistic missiles.

Although not part of SDI and even though a product of ten years of earlier work, the success of the Homing Overlay Experiment, conducted on June 10, 1984, showed the Europeans that there was, indeed, some progress in technologies needed for missile defense. Managed by the US Army's Ballistic Missile Defense Systems Command, this experiment demonstrated the non-nuclear intercept of one missile by another missile—hitting a bullet with a bullet, as it came to be described.

In July 1984, Lieutenant General Abrahamson made the first of many visits to NATO in an attempt to explicate the US approach. His enthusiastic, fascinating, and confident presentations to national delegations at NATO further bolstered the impression of US seriousness about SDI. Part of his message in press conferences at NATO was that what the Europeans had to worry about was not US deployment but the Soviet build-up of both offensive and defensive nuclear forces. After all, Soviet deployments had their own march and did not depend on the pace of US research efforts. Secretary of Defense Caspar Weinberger had made the same points at the Nuclear Planning Group ministerial meeting at Cesme, Turkey, in March 1984. And a White House pamphlet in January 1984 had documented Soviet efforts—including the radar construction at Krasnoyarsk.

General Abrahamson reassured the allies that there would be no rush to deployment; there would be time before such a decision to know whether the defensive technologies would live up to their promise or not, and there would be plenty of opportunities for the West to think through the implications. At the same time, there was urgency in the research, driven by concern with American vulnerabilty and by the need to ensure a survivable retaliatory capability.

Europeans throughout 1984 and 1985 gradually learned more and more about the SDI technologies (of particular interest to industrialists and scientists). There were many opportunities to plumb US thinking about strategic defenses, drawing on exchanges with Ambassador Nitze, Assistant Secretary Perle, General Abrahamson, and the US negotiators from Geneva.[12] Most of the exchanges served to elaborate the "four points" Mrs. Thatcher and President Reagan agreed in December 1984.[13]

At the Nuclear Planning Group (NPG) ministerial meeting on March 18, 1985 (some six days after the United States and the Soviet Union began the nuclear and space talks in Geneva), the United States invited eighteen Allied nations to participate in the SDI program. Besides the initial flap over the supposed ultimatum of a US deadline for a response, some nations privately greeted the invitation with consternation because now for the first time they had to do something about this *American* program. With great expectations, other nations began to look seriously at possible technological and economic benefits from participation.[14] A few nations hoped to establish a common response from Europe—perhaps through the Western European Union (WEU), the Euro-Group, or even the Independent European Programme Group (General Rogers' suggestion at one point).

Some suspected, especially defense experts and commentators critical of the initiative, that the United States was trying to buy European support for SDI with the invitation to participate. The potential competition between SDI and Eureka (a high-tech program sponsored by France) began to be played up in the European press and for a while began to take on the character of a test of commitment either to Europe or to the United States. Fortunately, the United States did not respond to this French contrivance. Instead, the United States declared publicly that there was no competition between the two programs and welcomed the Eureka for what it might eventually contribute to strengthening the economic well-being of Europe and thereby the security of the Alliance.

After the allies began to win a few SDI contracts, pressure built in the United States to cut off participation; Senator Glenn, leading the movement, drafted an amendment that would allow contracting to allies only when the work could not be done within the United States. This compaign to restrict Allied roles was part of the "Buy-America" reactions of the Congress to the negative balance of trade with several major allies. Moreover, US industry began to question what the allies might contribute to the research efforts that could not be done in the United States with US taxpayers' funding.

Some of the allies began to look for ways other than participation to keep the US efforts on strategic defenses manageable from a European viewpoint. The Reagan-Thatcher "four points," along with somewhat similar language agreed between Chancellor Kohl and President Reagan, gave the Europeans some assurance that the Americans would not rush off to deploy strategic defenses. At the same time, there appeared to be nagging anxiety in Europe that the United States bore watching.

In a controversial speech in early 1985, Sir Geoffrey Howe raised a number of SDI strategy and technology issues that needed additional Alliance study. It would beggar the imagination to believe that his speech about President Reagan's personal program would not have been approved in advance by the prime minister. Lord Carrington was also busy during these months, first in forging an Alliance approach to SDI and then in keeping it intact.[15] Lord Carrington's basic approach was to recognize the extensive Soviet efforts in strategic defenses and to assert that the West was acting prudently in hedging against a Soviet breakout from the ABM Treaty, as well as a Soviet breakthrough in the technologies needed to defend against ballistic missiles.

One way to ensure that an issue gets enough "chewing over" before any action is taken at NATO is for nations to initiate a study or, better yet, a series of studies. The spring of 1985 saw the first call for a study of SDI's implications for NATO's strategy. In its East-West Study of 1984, the Alliance had reaffirmed the "Harmel Report" of 1967, which provided the political underpinnings for relations, including arms control, with the East and provided a context for NATO's adoption of the strategy of flexible response and forward defense. In sum, the East-West Study held that NATO had its strategy just about right for dealing with the Soviets in regard to arms control, defense, and deterrence. Yet, the leader of the Alliance, the president of the United

States, continued to suggest that the strategy might not be right. This inconsistency worried a number of nations; by 1985 some defense experts were suggesting that NATO take another look, figuring in a number of factors including the potential contributions of strategic defenses.[16]

Added to other concerns was the unease in Europe over the "broad" interpretation of the ABM Treaty that the administration had begun to talk about in the fall of 1985.[17] The allies looked to the ABM Treaty as the most important hold they had on any US decisions to deploy strategic defenses. Mrs. Thatcher had secured agreement that the SDI program would remain compliant with the ABM Treaty, and the SDI program itself, in fact, had been so structured even before that agreement. In the UK's view, the ABM Treaty commonly understood, clearly not the new interpretation, prohibited development and deployment of any ABM system other than the fixed, land-based system at the one site allowed for each superpower. By late 1985, however, senior US officials were asserting that ABM components based on "other physical principles"—that is, on principles other than those known at the time of the treaty negotiation—could be developed and tested not only on earth but also in space. The treaty prohibited *only deployment* beyond the limited terminal defenses.

Among the original US negotiators of the ABM Treaty, only Ambassador Nitze defended the broad interpretation as the common understanding when the treaty was negotiated in the early 1970s. There was, of course, a clamorous reaction in Europe, as well as in the United States, to the new interpretation. After another intervention by Prime Minister

Thatcher with President Reagan, the allies secured an agreement, delivered by Secretary Shultz in San Francisco at a North Atlantic Assembly meeting (October 14, 1985), that the United States would abide by the so-called narrow interpretation at this time. The administration also asserted that the broad interpretation was justifiable. Since the SDI research and experiments had already been structured to be compliant with the treaty, there was no need to change anything.

The administration, in other words, did not back away from the broad interpretation but stated it was unnecessary to restructure the initiative. Since SDI would not be hampered by compliance with the treaty, the allies wondered what was the rush to test systems outside the common understanding of what the treaty allowed. Without the treaty narrowly interpreted, there would not be as clear a "firebreak" to demarcate research from the full-scale development and deployment of strategic defenses against ballistic missiles.[18] And if the superpowers were to deploy territorial strategic defenses, there would be much more pressure on Alliance nations to improve their conventional forces in order to deter war.

Apart from any potential consequences of SDI, the member nations had already decided to increase NATO's defense capabilities. In December of 1984, at West German and US initiative, NATO launched the Conventional Defense Improvements (CDI) special effort, with pledges from defense ministers to press for increased funding for defense, with determination to use resources more efficiently, searching for opportunities to increase defense capabilities in smarter ways—for example, through increased armaments cooperation. All of these measures sought to correct the areas of critical military deficiency identified by NATO's military authorities and agreed on by the political side of NATO's house.[19]

Some of the Europeans pointed to the SDI program to suggest politely to the United States that the constrained defense budgets of many nations, including that of the United States, would not be able both to pay for these conventional improvements and to fund deployment of expensive new systems for defense against ballistic missiles. In other words, the United States was put on notice that there would be no new defense funding available for strategic defenses. Since the United States did not seek funding from its allies for the SDI research (a few nations did fund some parts of the projects they were involved in), the allies had less claim to share control over the direction of the US program.

Except for the ABM Treaty debate, this period provided good opportunities for the allies to broaden their knowledge of SDI. The allies came to understand that the technologies would not be available for many years; deployment of anything like a "thoroughly reliable" system of defenses was decades away; the defenses being generally discussed for early deployment were not as revolutionary as had first been touted; and for a very long time, strategic defenses might protect not populations but military assets, leadership nodes, C^3, and other sites to be preferentially defended.

The number, quality, and level of Alliance consultations—bilaterally, multilaterally, and at NATO headquarters—improved greatly. With measured and frequent US explications of the research, a consensus emerged in Europe that the research was a prudent hedge and insurance for the Alliance, given Soviet activities. Europe wanted neither to be in the program nor out of it; the United States made it easier for the hesitant nations by not pressuring them for participation.

Europe's turnabout did not just happen. A very active US public diplomacy program began in earnest in January 1985. These extensive efforts included distribution of White House pamphlets on SDI and on the Soviet strategic defense programs,[20] the sustained involvement by US ambassadors from the major capitals, and the extensive work of Ambassador Abshire in Brussels, who served as the administration's leading official in Europe. The United States actively searched for the right audiences for senior US officials to address the character and pace of the research and made detailed and rapid response to news accounts and opinion pieces that misrepresented the SDI program.[21]

By mid-1985, supporters of SDI also included a number of influential members of the political elite, retired military leaders, and former government officials in the major NATO capitals, as well as important scientists, industrialists, and well-known political scientists in the think-tanks, study centers, and universities of Europe. Their argumentation aimed at European concerns helped to offset the negative critiques of powerful British defense experts such as Denis Healey and Lawrence Freedman. The latter had warned that Europe should wake up: the United States was gradually decoupling from Europe, and there were no SDI-derived strategic defenses—beyond what is now allowable in the ABM Treaty—that would be in Europe's interests.[22]

Either or both of two major decisions could have cracked the Alliance consensus apart during this period: namely, a US decision to deploy defenses (especially if any part were space-based) and the US unilateral abrogation of the ABM Treaty. A principal worry in Europe was that there would not be a "firebreak" between the research and the decision to deploy at least some systems, even if they were not fully effective. Ambassador Nitze's criteria of survivability and cost effectiveness at the margin—announced to the Philadelphia World Affairs Council in February 20, 1985—proved inadequate comfort in the longer term.[23]

Phase 3—November/December 1985 to Fall 1986. The period of late 1985 to fall 1986 began with the tamping down of the US "narrow" interpretation of the ABM Treaty and the first agreement on participation in the SDI program. Moreover, this phase also witnessed the start of discussions at NATO on defenses against tactical ballistic missiles (ATBM) within the German-inspired context of extended air defense (EAD) improvements, as well as successes in SDI research itself—technological developments that encouraged those wishing to deploy. Late in the period, however, was the bittersweet of the Reykjavik summit. SDI ironically stood in the way (for all the wrong reasons) of far more disastrous agreements, from a European viewpoint, between the United States and the Soviet Union on the elimination of ballistic missiles and even on the desirability of a world without nuclear weapons.

1. Allied participation. Several of the major nations agreed to participate government-to-government, from late 1985 to the fall of 1986. On December 6, 1985, the United Kingdom became the first to sign a Memorandum of Understanding with the United States on participation. The Federal Republic of Germany reached agreement with the United States on March 27, 1986, with the negotiations handled by the FRG Economics Ministry, not the Defense Ministry. These agreements were followed in 1986 by one with Israel in May and with Italy in September.[24]

A number of nations followed the Norwegian model of response: namely, no government-to-government agreement on cooperation, but companies were free to bid for contracts. France had taken a similar position in 1985, but France had also been critical of SDI over strategy issues (i.e., as a threat to France's independent deterrent), as well as on economic grounds. France saw SDI as a means through which the United States would leap so far ahead in the next generation of technologies that Europe would not be able to compete; the United States would also be draining the best "brains" out of Europe. Once Prime Minister Jacques Chirac entered the "cohabitation arrangement" with President Mitterrand, however, France's reactions to SDI changed. In fact, Chirac stated that "France cannot afford not to be associated with this great research programme."[25]

The smaller or less industrialized nations such as Portugal, Luxembourg, Spain, and Turkey let it be known that while they did not reject the US invitation, they had little to offer and had no expectations of any participation.

In sum, despite European misgivings about the value of strategic defenses and about how deployment would affect deterrence and arms control, no NATO country actively worked against SDI, at least in the open. As is evident in the footnotes to NATO communiques in this period, particularly those from the Defense Planning Committee (DPC) and the Nuclear Planning Group (NPG), Greece and Denmark, as well as occasionally Norway (depending on the formulation of the language), often "took footnotes" expressing reservations about SDI. While NATO was used to Greek and Danish reservations on nuclear matters, a footnote from Norway was far more worrisome a development. As time went on, the communiques ceased to be litmus tests of solidarity in regard to SDI, except, of course, for support of US negotiating positions in Geneva.

2. *The anti-tactical ballistic missile.* The introduction of anti-tactical ballistic missile (ATBM) efforts into NATO during the winter and spring of 1986 paralleled SDI research into architectures needed in theaters of potential war, especially Europe and to a lesser extent Korea, Japan, and the Middle East. The SDIO work, along with the US Army's discrete efforts into defenses against tactical ballistic missiles, emphasized architecture and battle management concepts.

On both sides of the Atlantic, some defense experts suggested that the substantial benefits from the research would be at the theater (not the global level) in the defense of military assets (not populations). The experts also held that the president's vision of a global defense against ballistic missiles would remain a fanciful excursion—and a dangerous one at that. There was, in other words, no escape from the existence and terrible destructiveness of nuclear weapons, and no responsible leader should offer any hope of escape. In the view of the experts, theater defenses could be available much sooner than the more exotic technologies; such defenses would help with immediate threats: e.g., the shorter range, new Soviet tactical ballistic missiles that threatened critical defense assets.

James Woolsey, for example, in the fall of 1985, suggested that European worries about the United States retreating to a Fortress America through SDI—Charles Krauthammer later termed this idea as US yearning for "impatient isolationism"—could be "assuaged if the US and its Allies were to make a serious effort, together, to develop tactical ballistic missile defenses for NATO." [26] The Fletcher study had already made this suggestion two years earlier: namely, that some of the technologies for terminal defenses, ground based, would be worth deploying while investigation of the more advanced technologies were under way. Woolsey saw no conflict between a NATO ATBM effort and SDI in that some of the same technology, applicable to defense against SS–12s and SS–23s, could also be used against SLBMs and ICBMs—for example, airborne optical sensors and kinetic energy interceptors.

The credit for getting the Alliance to take a look at the threat from Soviet TBMs and at possible NATO countermeasures belongs to then German Minister of Defense Manfred Woerner. From the first presentation of his ideas in 1985 and later at the Wehrkunde Conference in March 1986, Woerner insisted that efforts to develop an anti-missile system should remain apart from the US research in SDI.[27] At the same time, he understood that NATO might draw from SDI research on global systems. The United States supported Woerner publicly, including the preference for keeping ATBM efforts discrete. The US position was to treat the ATBM program as separate from but consistent in objectives with SDI research.

At the outset of ATBM discussions, some leaders suggested that there be a European Defense Initiative (EDI) in parallel with the US SDI. [28] However, on the margins of the NPG ministerial meeting in the spring of 1986, Woerner made it clear that he did not accept any EDI concept. Instead, NATO should put its efforts under an Extended Air Defense (EAD) initiative; NATO needed to go ahead with this program even if there never would be an SDI deployment. Woerner quite rightly believed that dissociation from SDI would mean less political baggage for the effort. Within the US Congress, a number of members pressed for ATBM/ATM efforts in the belief that such defenses could take advantage of the ground-based technologies for terminal defenses that had come a long way from the mid-1960s.

3. Early deployment. The discussion of theater defenses in NATO fora[29] had a parallel hearing among those in the Congress and in the administration who wished to move rapidly toward early deployment. For some advocates, hardware became the idol—never mind waiting for the research. While there had been progress in the technologies for terminal defenses, ground-based defense of military assets had never been the prime focus of SDI. All of these and other elements led to debate on deployment in the United States that was contentious, but discussion did not approach the explosive character it had in some parts of Europe.

The Alliance had supported SDI based on the assumptions that there would be a "firebreak" between the research and any decision to deploy and that such a decision would not be made until the early 1990s. By mid-1985, Europe had reached the conclusion that there was nothing to worry about; the United States was committed to consultations and to negotiations before taking any decision to deploy systems. The United States had committed not to proceed with deployment until there were answers to questions the president had raised in 1983, and this would not happen until the 1990s. Yet in the fall of 1986, senior American officials were publicly discussing early deployment without any particular cognizance of or sensitivity to European views.

The United States had done well in consultations through 1985 and most of 1986, but by the fall Europeans began to have doubts about the US commitment. They clung tightly to the requirement that there would have to be some accommodation with the ABM Treaty before any deployments—either through negotiation of changes to allow additional deployments or through abrogation of the treaty. Yet on this point the United States had already committed itself, averring there would not be any need to restructure the SDI program along the lines of the "legally correct interpretation" of the ABM Treaty.[30] Some of the allies thought they had been had. For if the United States moved to early deployment of even the limited, terminal defenses in the near term (the next five to seven years) as allowed by the treaty, then the die would have been cast in favor of strategic defenses even before the Alliance had a view on the policy consequences of strategic defenses.

Some SDI supporters in the United States were eager to get systems deployed to show the Congress some results of the monies spent to date, especially after it became clear that SDI would not obtain the funding envisioned at the outset of the program. Many supporters looked to evolutionary systems. One of the difficulties with this approach, however, was that the Soviets with their warm production lines for terminal defense launchers and interceptors would be able to move more quickly than the United States in getting systems in the field.[31] A Rand study in the fall of 1986, for example, had concluded that if the United States did not redress existing asymmetries with the Soviets in ballistic missile force capability through modernization of basing modes or through arms control, "it must prepare to build and deploy strategic ballistic missile defense capability nearly twice as fast as the Soviet Union builds and deploys its strategic defenses." [32]

Although most of the pressure in the fall of 1986 came from SDI supporters in the Congress (e.g., Congressmen Kemp and Courter), the same proposals in the administration gained currency again in the late winter and early spring of 1987. In a speech to the Commonwealth Club, for example, Secretary Weinberger said that the rapid progress made in technology research "convinces us that a phased deployment of strategic defenses, which could begin as early as 1993 or 1994, is the most feasible way to reach the president's goal of a thoroughly reliable defense."[33]

Weinberger was optimistic even with the loss of the Shuttle a year earlier, asserting that a deployed first phase of SDI was "one of the best ways to motivate arms reductions" and that such defenses would tell the Soviets "no first strike." As Richard Perle had forcefully argued the point with the allies on several occasions, Weinberger made the case that the United States was intent on devaluing the worth of ballistic missiles through strategic defenses. In his view, "Our demonstrated will to begin to deploy strategic defenses will indicate to the Soviets that the offensive ballistic missile is on the path to ultimate extinction"; furthermore, "... a first phase of strategic defense would devalue these nuclear missiles and complicate their use." This part of the explanation remained true to the original "vision" of rendering ballistic missiles "impotent and obsolete."

Weinberger tried to show how phased deployment fit into the president's vision as "simply the practical means of moving toward a defense of our nation and that of our allies. And it is the defense of our nation and that of our allies, not the defense of missile silos or some other specific target, that is our goal. Each phase, including the first, would be a part of and contribute to, the entire system."[34] Weinberger set out the agenda of the administration for dealing with ballistic missiles—deep reductions as the "centerpiece" of arms control and then elimination of the "military value and first-strike potential of ballistic missiles through the deployment of strategic defenses." This was precisely the approach the allies did not want: a foregone conclusion that deployment would be valuable even before the results were clear.

While Weinberger believed that "common sense and strategy tell us phased deployment is the way to proceed," the Europeans did not share his version of common sense and strategy. Quiet did not return to the Alliance on the issue until Secretary Shultz, along with Secretary Weinberger, finally affirmed that there would be no decision on early deployment in the Reagan administration. From the fall of 1986 on, Senator Nunn led those in Congress who opposed population defense as fanciful at best and as a dangerous diversion from arms control and force modernization.

4. The Reykjavik summit. Much has already been written about the Reykjavik summit of October 1986, with some few commentators treating the meeting as an important crossroads in the history of nuclear weapons, deterrence, and arms control. But most others have treated it as a debacle that needlessly jeopardized the security of the United States and its allies through dalliance of the presidential group with arms control positions that had not been well thought through. One of the ironies about Reykjavik for the allies was that it was the president's refusal to give any ground on restricting SDI research to the laboratory that prevented, rightly so from the European viewpoint, any agreement to eliminate ballistic missiles over a ten-year period.

Within a few days of the summit, Mrs. Thatcher made another trip to the United States to secure agreement on a common approach the allies could take to arms control. In the European viewpoint, considerable confusion had followed the summit concerning what had been agreed and even what had been discussed. The confusion in the use of the terms "ballistic missiles" and "nuclear weapons" in press conferences following the summit was particularly egregious in that some of the US spokesmen appeared to make no distinctions at all. The complaint heard for weeks after the summit was that this toying with European security without adequate consultation was another example of the lack of US respect for allies and evidence that the United States was looking out primarily for itself.

Phase 4—December 1986 to December 1988.
Europe gradually became less hysterical about what SDI was and where it was going. While in 1985 a minimalist consensus on the research had obtained in Europe, the allies now cherished the further understanding that SDI would not mean deployed systems any time soon. Despite any technological breakthroughs and despite the desire of some administration officials to press for early deployment, the Europeans recognized that nothing would jeopardize the consensus on the research. Four major factors helped to calm European anxieties.

First, the pressures on the US defense budget because of growing federal deficits and the need to reduce the national debt were becoming enormous. Europeans knew that in the context of the Gramm-Rudman-Hollings legislation, there simply would not be sufficient growth in funding for SDI to allow early deployment even if that decision were to be taken unilaterally. The United States would also not be in a position to pressure Europeans on their defense spending, and the funding would not be available to pay for the increased conventional force capability that would be required if deterrence were based more and more on defensive systems.

Second, in the US election year, neither political party could afford to commit to deployment of strategic defenses when the results of the SDI research were still not known. No candidate would campaign on the need for large funding increases targeted for strategic defenses when the basic defense budget was already at negative real growth. Moreover, as is usually the case, domestic issues captured interest of the politicians and the public.

Third, INF deployment and arms control were center stage in Europe and in the United States. With agreement on INF missiles reached in late 1987 and with the ratification process under way in 1988, there would be no possibility for the United States to launch consultations with the allies as a preliminary to abrogation of the ABM Treaty; no opportunity to conduct experiments that would push the edges of knowledge about defenses in space; and no chance to press for early deployment when Senator Nunn was using the INF ratification hearings to lock in administration acceptance of the ratification record of the ABM Treaty.

Fourth, the potential and promise for radical changes in the West's relationship with the Soviet Union loomed, now with Gorbachev in charge. He was a force to be reckoned with, a man whom Mrs. Thatcher identified early on as someone "we can do business" with. In the view of some in the Alliance, Europe needed to help the United States deal with Gorbachev, especially since he had nearly "tricked" the United States into accepting its own offer on elimination of ballistic missiles.

With Reagan's insistence that SDI would not be a bargaining chip and with Gorbachev's insistence that the ABM Treaty precluded development and testing of space-based strategic defenses, there was little chance of any breakthrough on these issues at Geneva. Once the Soviets agreed to break the linkage between SDI and INF, SDI would be tied only to the Strategic Arms Reduction Talks (START) reductions—a negotiation primarily the business of the two superpowers. As long as the ABM Treaty stayed intact and as long as the superpowers refrained from testing strategic defense components in space, the UK and France would not have to worry about Soviet strategic defenses rendering futile their own expensive and controversial (at least in the United Kingdom) modernization of nuclear ballistic missiles.

MAJOR CONCERNS RARELY EXAMINED

The European discussion on the Strategic Defense Initiative (SDI) has been conducted in terms of technological and industrial policy, Alliance policy, and more recently, the benefits of the strategic defense research for enhanced air defense in Western Europe. The more basic strategic issues which result from the transition to a defense-dominated world, which President Reagan and some of his advisors would prefer, have rarely been tackled.[35]

But for the work of Glenn Kent, Randall DeValk, and James Thomson of Rand, as well as some classified work in OSD, in the JCS, and at SHAPE, there have been few serious attempts at formal study of the implications of strategic defenses for NATO's strategy or of a workable transition to strategic defenses. In the academic community, Colin Gray, Keith Payne, Arnold Kanter and Albert Wohlstetter[36] asked many of the right questions about SDI and strategy, about the transition to deterrence based more and more on strategic defenses, and about the post-transition period when conventional armed forces would be far more important as a principal deterrent to war in Europe.[37]

Although SDI has been the focus of innumerable commentaries, the aspects that have been only rarely discussed include the potential consequences of Soviet strategic defenses for NATO's strategy and for the penetrativity of French and British ballistic missiles (perhaps for obvious reasons), as well as the implications of Reykjavik discussions on eliminating ballistic missiles. Europe appeared to believe it would not have to deal with strategic defenses at all, were it not for the United States. The admission by General Secretary Gorbachev in Washington at the December 1987 summit that the Soviet Union was well along with many of the same technologies that the United States has under study in SDI cannot be totally dismissed as braggadocio—especially since the United States had been trying to get the Soviets to admit the existence of their extensive program for many years.

DETERRENCE STRATEGY REVISITED

President Reagan's challenge to find a means other than assured destruction to underpin the strategic relationship struck a blow at NATO's strategic concept. For no matter how well US spokesmen might have argued the case, there would not be a coincidence of shared objectives between the new vision and NATO's concept: to defend populations and not military assets alone; to deny the enemy his objectives rather than to deter by threatening unacceptable damage in retaliation; and to devalue ballistic missiles until they are rendered "impotent and obsolete" rather than to remain vulnerable for the sake of assured destruction by survivable retaliatory forces. Reconciliation between this version of SDI and NATO's strategic concepts seemed impossible.

However, the SDI presented to the allies by the United States after January 1985 was quite different in essentials. The near-term emphasis was on protecting military assets; SDI was now said "to enhance deterrence" rather than to be a new strategic concept. There would be a mix of offensive and defensive systems at the end of the transition, not a ridding the world of the threat of nuclear weapons (the Reykjavik summit notwithstanding).[38] With the exception of the SDI Organization itself and other "believers" in the vision, the bureaucracies had turned SDI into an evolutionary program. The allies shook their heads in confusion over which SDI to respond to. The basic goals of SDI seemed to be amorphous.

What the allies found missing was any analysis of what the United States apparently took for granted: namely, assuming the technologies would be available for effective strategic defenses, deterrence based on such defensive systems would be better and more stable than deterrence based on offensive retaliation.[39] On this side of the Atlantic, even among the academics, there was some expectation that SDI would encourage an overdue reexamination of NATO strategy and doctrine.[40] Preliminary discussion of the study idea at NATO included an examination of the offense and defense, at US insistence. With such a focus, the Alliance would have to look at SDI in the context of arms control, of nuclear modernization on the part of the West, and of Soviet offensive and defensive efforts.

Offensive Capability of Strategic Defenses. A number of the allies questioned whether deployed strategic defenses, especially components in space, might not add as much capability (or even more) for the offense as for the defense. Weapons in space, whether kinetic interceptors, radio devices, or directed energy beams, could be used in an antisatellite role to pluck out Soviet "eyes" needed for the command and control of nuclear forces, warning and attack assessment, battle management of strategic defenses, and reconnaissance and surveillance as well as verification of arms control agreements.

Unfortunately, serious discussion of this offensive capability issue is often encumbered with fictitious stipulations and imaginative speculations about the use of weapons in space against targets on earth. Such use is most often physically impossible, or militarily insignificant, or prohibitively expensive. Still other critics have pointed out that with a decision to deploy strategic defense weapons in space, the superpower could not then ever achieve a comprehensive treaty banning nuclear testing, nor could an ASAT Treaty ever be negotiated, at least for weapons that could attack satellites in low earth orbit. There is little prospect of distinguishing what is offense and what is defense in space-based weapons and battle management capabilities since the distinction could rest as much with intentions as with capabilities. Pre-emption against the enemy's C^3 in space, for example, because an attack appears imminent could obviously be a defensive strategy using defensive and offensive weapons.

In the original 1983 speech, the president himself recognized that defensive systems "have limitations and raise certain problems and ambiguities." He continued, "if paired with offensive systems, they can be viewed as fostering an aggressive policy, and no one wants that." The obverse of this proposition is also compelling: namely, the West could not take the chance that the Soviets alone would have effective strategic defenses to be coupled with their first-strike capabilities.

Technology Driving Policy. One point often repeated in Europe was that the Americans are at it again, the old Yankee desire to make the good better and to tinker with things not broken. Given the modernization programs under way in the United States and in Europe, both conventional and nuclear, and given the arms control positions of the West, European audiences asked whether there was any problem with deterrence in the first place.

The debate includes the false disjunctive of either "policy driving technology" or "technology driving policy." The reality, maybe even the ideal, is that there is and should be interaction between policy and technology. One of the persuasive arguments, at least in some US quarters, about launching SDI research was that there were promising technologies available. These technologies were not at hand in 1972–74 when the superpowers agreed to forgo strategic defenses other than at one fixed land-based site. Policy should change to match the technologies available, as some would have it.

The Europeans apparently assumed that strategic defenses against ballistic missiles, forever after the ABM Treaty, would be shunned by the signatories. However, neither the Soviets nor the United States ever accepted that proposition so absolutely stated; nor did either party accept the notion of vulnerability in perpetuity. Despite not being signatories, NATO European nations presented their views repeatedly on the importance of continued adherence to the treaty.

Many European industrialists have taken a more favorable stand toward SDI than have politicians and defense experts in and out of government. The industrialists at conferences throughout the major nations made telling points in the European debate, saying the competition in offensive and defensive armaments will continue; work on lasers, particle beams, C^3, and kinetic energy weapons can strengthen deterrence even if strategic defenses are never deployed; and the Soviets and the West were at work on many of these technologies before there was an SDI. No matter what the politicians and the defense experts thought about strategic defenses, if the Soviets alone had effective defenses against ballistic missiles in place, deterrence and not just ballistic missiles would be rendered "impotent and obsolete."

Some industrialists and policy-makers on both sides of the Atlantic began in mid-1985 to make much of the potential for spin-offs from the SDI research to conventional military capabilities, especially C^3, and to civilian applications. The US Government spokesmen remained wary of this line of reasoning, not because there would not be substantial "translations" of the research into other areas, but because the United States should not even appear to be "buying" the support of the allies through the attraction of spin-offs.

Some in Europe, with the then SACEUR General Rogers among them, saw the SDI as little more than an endless series of research projects that would draw off needed funding from much more important improvements in conventional forces, without improving the military balance. Others saw

SDI as a way for the US to leap ahead technologically. French Foreign Minister Claude Cheysson, for example, asserted, "In the name of the threat which they claim hangs over the United States and Europe, it will be possible to inject considerable sums in scientific and technological research. The Americans," he continued, "wish in this way to recover their leadership in certain areas of the high technology of tomorrow."[41] As Glyn Ford (member of the European Parliament, UK) explained in his 1987 article, "The Dangers of SDI," "While the European governments disagree with the SDI in political and military terms, they see involvement in SDI as a means of access to American advanced technology and an opportunity they cannot afford to miss."[42]

The larger nations also viewed participation in the SDI program as a way to have some control over the direction of the research and, indeed, over the decisions on full-scale engineering development and deployment. While not signing up to this approach, US spokesmen did suggest that those nations participating in the program would be much more knowledgeable at the time of later decisions and that their voices for that reason might carry more weight.

Two sub-arguments about technology were prevalent in Europe from mid-1985 through late 1987: that SDI technologies would not work because "perfect defenses" could never be achieved and that the SDI systems could not be adequately tested to know whether or not they would work.[43] While no government took these positions officially, a number of European defense experts and scientists lent their voices to similar arguments being made in the United States by the Union of Concerned Scientists and other critical groups (including some in the Pentagon). Many of the examples centered on the extraordinary demands strategic defenses would place on battle management and C^3 capabilities and on the idea that the strategic defensive system would have to work near perfectly the first time—with no realistic testing in peacetime.

The argument eventually hinges on what objectives have been set for the defenses and how well strategic defense systems deployed incrementally might be able to accomplish the objectives. It would seem, however, that every increment of defense would serve to devalue deployed ballistic missiles and possibly to undercut the rationale for increased numbers of missiles. Moreover, the testing that could be done along the way would be on the various parts of defenses and their integration over time; there would appear to be no reason why such testing of the defense could not be at least as rigorous as that of the offense—the establishment of the National Test Bed in Colorado Springs responds in part to the need for testing components, systems, and systems of systems. Of course, no testing could prove the whole system would work in every particular.

The Arms Race in Space. Perhaps more than any other single concern about SDI, there was a clear presumption in Europe against any deployment of arms in space, whether Soviet or US arms. Except for the security elite, there appeared to be little recognition in Europe about how much military assets in space already contribute to US and NATO capabilities: in national technical means for verification of arms control agreements, surveillance, navigation, tracking, weather, battle management, and communications capabilities.

The fact that the Soviets and the United States have been using space for military purposes since 1958 when the Soviets tested the first ballistic missiles does not argue for increasing the military use of space, especially placing arms in space. Even though the Soviets for a long time have had an ASAT weapon and have tested it in space, and even though the Soviets have a deployed ABM system around Moscow (including exoatmospheric, nuclear-tipped Galosh interceptors), the Europeans wish to preclude any further erosion in the use of space for weapons deployment. Despite the Soviet capabilities, for example, Muller strongly cautioned that since "space-based strategic defense weapons would inevitably contain some ASAT capability, this is an additional reason for Europeans not to embrace the SDI concept too enthusiastically."[44]

The European Space Agency, the European Parliament, and the European Community have all stressed the need for "peaceful" commercialization of space. When France posed Eureka as an alternative to the US SDI program, the argument was that the Europeans could work on peaceful technologies for the benefit of mankind while the United States would be working on technologies for military purposes. It was not too long into the Eureka program, however, before France was talking about potential military applications of some of the technologies under investigation.

Morality of Nuclear Deterrence. In his initial SDI speech, President Reagan said that he had become "more and more deeply convinced that the human spirit must be capable of rising above dealing with other nations and human beings by threatening their existence." He went on to say that the specter of retaliation and of mutual threat is "a sad commentary on the human condition." European leaders were totally opposed to any suggestions from the Americans that deterrence based on the threat of massive retaliation, which had kept the peace for more than forty years, was now seen to be less than satisfactory in its morality.

Even before the indirect criticism of deterrence by the US president, the Europeans had been struggling with peace movements and with religious organizations that had questioned the morality of deterrence. In fact the Roman Catholic bishops of the United States, in their pastoral letter of May 1983, had narrowly found "deterrence based on balance" to be "morally acceptable," just as Pope John Paul II had in June 1982 in his speech to the UN Second Special Session on Disarmament.

The task fell to Ambassador Nitze (special advisor to the president and secretary of state on arms control matters) in the late fall of 1985 to reassure the Europeans on the issue of the morality of deterrence. In a speech entitled "SDI: An Ethical Evaluation," Nitze presented the Institute for Theology and Peace in Bonn, West Germany, a spirited explanation on how "deterrence based on the ability to defend [not just to retaliate] is moral":

> If there is no available alternative other than the threat of nuclear retaliation, then this is the necessary and moral course. If, through adequate defenses, as envisioned by the SDI concept, one can deter attack primarily by denying a potential aggressor the prospect of military success, movement toward greater reliance on defenses becomes the preferable and the moral course.[45]

Apart from any judgment on the ethical and theological value of Ambassador Nitze's position, the point is that the Europeans were enough upset by the morality issue to have needed this sort of reassurance from the US administration. The United States found the present strategy of nuclear deterrence, absent any alternative, to be a moral strategy.

A Unified European Response? Some of the nations, such as The Netherlands and Italy, looked at the Western European Union (WEU) as a possible forum in which to take a common approach on SDI to the United States. As we discussed earlier in this chapter, none of these calls for unity ever came to much except a few minor studies and informal discussions among the allies, especially in advance of major NATO meetings.

That said, however, even in the piecemeal way that individual nations dealt with the United States on SDI issues, a cumulative set of informal guidelines gradually did evolve. The European nations hoped the United States would adhere to them in continuing the SDI program:

—Not to call the strategy of flexible response into question before there is a reasonable alternative strategy.
—Not to spur an offense and defense arms race with the Soviet Union through SDI.
—To think carefully about the leverage SDI has at the Geneva negotiations. Translation: use SDI as a bargaining chip for deep reductions in offensive forces. (Note: The Europeans knew the depth of presidential commitment to SDI and never suggested its abandonment

in exchange for arms reductions, at least in public.)
—To preserve the ABM Treaty as it is. This treaty is the firebreak between research and deployment.
—To keep the superpowers from doing anything that would threaten stability. For the *status quo* to obtain, it is necessary to have a United States vulnerable enough for coupling purposes but not so vulnerable that the United States questions its commitments.

Given trade deficits, budget deficits, the budget pressures of Gramm-Rudman-Hollings, the focus on the INF Treaty, and the election year in the United States, by late 1987 there was much less to worry about in regard to SDI.

3. SDI AT EUROPEAN CAPITALS AND NATO

Allies in Europe for the most part avoided occasions for publicly advising the United States on how to proceed with SDI. In private exchanges and at industrial conferences, however, Allied defense experts, as well as occasionally even government and NATO officials, were less reluctant to dispense guidance. This advice often was not entirely coincident with public refusal by most member nations to support anything more than SDI research.

MANAGING SDI IN NATO EUROPE

In the arms control consultations at NATO, advice about SDI was more frequent and generous than elsewhere, especially since many allies believed that success in arms control negotiations would depend on concessions from the United States. The Soviets, most thought, would never go ahead with START without first killing SDI. One of the reasons President Reagan had to repeat so often that SDI was not and would never become a "bargaining chip" was not so much for the Soviet Union but for the allies to understand.

Allied leaders and bureaucrats did not tire of reassuring their own publics that SDI was a US, not a NATO, program. Moreover, they emphasized that beyond research there was no commitment within the Alliance to ballistic missile defenses, whether strategic or tactical defenses. The NATO members could avoid entanglement in controversies surrounding the research and could point to SDI as a prudent hedge by the Alliance against Soviet developments. In taking this approach, the governments in power were taking their cues from public opinion in Europe; the one exception was Prime Minister Thatcher, who tried to mold British understanding of SDI research. On the whole, European opinion remained negative about strategic defenses, a view at odds with broad public and congressional support for SDI research in the United States, especially in the first three years of the Reagan program.

Since SDI was so closely associated with President Reagan, most European NATO governments, in public at least, did not want to give offense by criticizing his initiative directly. Even Prime Minister Thatcher was careful, in winning Reagan's approval for her arms control agenda, not to reject the concept of strategic defenses. At home, however, some of her senior diplomats were less circumspect in pointing out the difficulties SDI posed for NATO's security arrangements and strategy, for arms reductions, and for the independent British nuclear deterrent. The lack of caution and the stridency of comments, indeed, were what was so startling about Foreign Secretary Geoffrey Howe's speech of March 1985. Howe pointed out strategy, technology, and arms control concerns about the SDI program, suggesting that the Soviets might cheaply overcome such a Maginot Line defense and underlining "the importance of proceeding with the utmost deliberation."[1]

To protect the president's initiative, then Assistant Secretary of Defense for International Security Policy Richard Perle led a rapid and heavy counterattack against Secretary Howe's views in the international press and in a number of British forums. Needless to say, the United Kingdom and the United States, given their "special relationship," presented an unusual spectacle, slugging it out in public over an issue at the core of the deterrence strategy of the Alliance. The two heads of state, Mrs. Thatcher and Mr. Reagan, remained agreeable and deferential friends while their hirelings, Perle and Howe, engaged in strategy-wrestling, with support for SDI in Europe the supposed prize. Howe wrestled, Perle "wrastled," but no prize was awarded.

After the initial endorsement in March 1985 of SDI as being "in NATO's security interest"—an endorsement the United States wrenched from its allies at the Luxembourg Nuclear Planning Group (NPG) meeting of defense ministers—NATO members did not again achieve unanimity on the progress and direction of the SDI program, even apart from "routine" difficulties with the French.[2] One or another nation, thereafter, would object to proposed and even final wording expressing support for SDI in communiques at ministerial meetings. That discord alone assured the allies that no further litmus test of support for SDI development and deployment was necessary during the remaining years of the Reagan administration.

At subsequent ministerial meetings, the United States acquiesced, to the relief of Alliance members, to less specific wording about SDI in communiques from foreign and defense ministers. The thrust of the new language, almost formulaic in diction with the full meaning known only to insiders, was that the allies supported US negotiating positions in Geneva arms control talks. Since US positions in the so-called third area at Geneva included the potential of strategic defenses among "space and defense issues," the United States could read the communiques as an endorsement of SDI research. At the same time, the allies could read the communiques as having little or nothing to do with SDI proper.[3] Even with this papering over of differences, however, two nations, sometimes three, repeatedly found it necessary, from the spring of 1985 on, to take exception through footnotes to the language on SDI in the communiques issued at the close of NATO ministerial meetings.[4]

Fortunately, the United States resisted the temptation to press for "solidarity" from the allies. The administration needed to play down Allied criticism of SDI, and the allies knew it. The American public was not fully aware of the pusillanimous support the allies offered. As a matter of course, in all issues Alliance members are judicious lest they petition the shibboleth of "solidarity" too often. By tacit agreement among the allies, a call upon "solidarity" is and should remain the last resort as an argument for unanimous consent. What the United States settled for was a fragile consensus among the nations that SDI research was a prudent hedge against parallel efforts of the Soviet Union and insurance against the possibility of a Soviet breakout from the ABM Treaty toward a territorial defense against ballistic missiles.

LORD CARRINGTON AND THE NATO STAFFS

From 1984 to mid-1988, Lord Carrington, secretary general of NATO, strove to hold together

Alliance support for SDI research. He took every opportunity to ensure that the United States continued comprehensive and frequent consultations in Brussels on the direction of SDI, along with bilateral and multilateral contacts in NATO capitals.[5] Carrington may well have feared that the United States would also unilaterally decide to follow the "broad" interpretation of the ABM Treaty and that without adequate consultations the United States would unilaterally decide to deploy early phases of space-based defenses. Although the allies were not signatories to the treaty, they nonetheless viewed the ABM Treaty as a way to cap Soviet and US efforts to develop strategic defenses.

Carrington also worried that a US decision to deploy defenses might occur before NATO could investigate and validate, if appropriate, the contribution of such defenses to strategic stability. His nightmare was that the United States would not respect the "firebreak" between the SDI research and the development leading to deployment, a concept he tried to reinforce at every turn.

Lord Carrington was sensitive to the implications of SDI, not only for NATO strategy, but also for arms reductions negotiations in Geneva and for the "unintended" contribution of SDI to the "denuclearization" of Europe. Because he met with President Reagan several times a year, Carrington knew the president would not change his commitment to SDI. Lord Carrington, therefore, did not accept the bargaining chip approach; he also understood that the Soviets were conducting research in many of the same technologies. He kept a wary eye on how SDI might, even inadvertently, damage chances for progress in Geneva. More adroitly than any other official in the Alliance, Carrington carefully balanced the desires of the United States and the desires of the European allies concerning strategic defenses. After all, his job required the building of consensus; he worked earnestly to hold that consensus together.

To his credit, Carrington found ways to manage SDI issues at NATO; he knew the European thesis—that the US homeland should remain vulnerable to ensure that extended deterrence remained credible to the Soviets—did not sit well in Washington. The American public would have found the point self-serving and probably, right or wrong, would have rejected such requirements for extended deterrence.

One of Carrington's accomplishments was to contain pressures by some Alliance nations for a formal analysis of SDI's implications. He arranged ways to vent European frustrations through periodic, informal discussions among the ambassadors at NATO. He understood why the United States and a number of other nations, each for its own reasons, adamantly refused to begin any NATO studies before at least the early results of SDI research were available. Senior officials on both sides of the Atlantic did recognize that at some time NATO would have to examine the implications of SDI for strategy. As US policy-makers insisted, however, such an examination would also have to include an assessment of the potential consequences of Soviet strategic defenses. Moreover, no NATO nation wanted such a study to open the entire strategy to scrutiny; they knew they might not be able to put the strategy back together again.[6]

At the North Atlantic Assembly meeting in San Francisco in October 1985, Ambassador Nitze (special adviser to the president and secretary of state on arms control matters) suggested that "with clear, objective, and open minds," the West should "consider the implications of greater reliance on defenses. Those implications will weigh heavily in any decision we might take to proceed to deploy new defensive systems."[7] Despite this suggestion by Nitze that SDI's implications needed "consideration," neither the United States nor most allies appeared willing to enter a process that might question the credibility and worth of NATO's strategy. It took the events of late 1989 and 1990 to do that. Informal discussions among the NATO ambassadors became a means to peel the onion-like layers of implications—slowly and tearfully indeed.

While several nations went to the discussions in 1986 and 1987 urgently asking how SDI related to NATO's strategy, they were less enthusiastic when it came time to decide that the strategy needed review. This was especially true if the strategy needed review not solely because of SDI but also because of other "difficulties."

Confusion Among Strategists. The new paradigm of strategic vision President Reagan had hoped to achieve continued to elude the scrutiny of strategists and defense analysts, both in the United States and in Europe. European defense experts like Lawrence Freedman looked upon the promises of SDI not as vision but as myopia on the part of American leaders who were tiring of their long-term commitments to a Europe now strong on its own terms.[8] For many of these critics, SDI was of a piece with the sometimes strident congressional calls for the withdrawal of substantial numbers of US forces from Europe and with complaints in the United States about allies who were not paying a fair share of the security burden.

On this side of the Atlantic, as in Europe, the vision proved every bit as opaque. Charles L. Glasser, in his article, "Why Even Good Defense May be Bad," put the case succinctly:

> Unfortunately, a world in which both superpowers deployed effective defense is far less attractive than its proponents suggest: even after making the most optimistic assumptions, defensive situations might not be more secure than assured destruction situations; and the more likely outcomes of deploying BMD would place the U.S. in a situation far less secure than today's.[9]

In the public debates in Europe and the United States, only those unalterably opposed to and those unalterably aligned with SDI claimed any surety in their views. For the rest, there remained a muddle of contradictory and ethereal distinctions, tentative approvals, and complex assessments—all too cluttered with conditions to be meaningful to the public.

Responses without Analyses and Evidence. As the strongest nation in the Alliance, the United States nearly always won the day-to-day exchanges on SDI in NATO forums. When the persuasiveness of argument failed, the weight of the United States did just fine. Many of the allies, though, were tuned into the relatively few but uniformly discouraging studies of the strategy implications and technology assessments of SDI, primarily from the United States: for example, the assessments and commentaries of the Office of Technology Assessment,[10] the Rand Corporation,[11] and the Union of Concerned Scientists,[12] as well as the occasional remarks from spokesmen or individual scientists at US laboratories (such as SANDIA and Lawrence Livermore). In short these studies (through 1988) concluded that, at best, reliable defenses were a long time away, if in fact they could ever be achieved.

The allies, especially Britain and West Germany, also closely followed congressional commentary from influential members. Senator Nunn, who was one of those closely watched by the allies, supported the research and some limited deployment of terminal defenses but did not make the leap of technological faith to full deployment of space-based systems and to population defense as the primary goal for SDI. From a European viewpoint, it would be a lot easier to live with an SDI directed towards strengthening deterrence than with an SDI dedicated to replacing the current strategic concept with a more defense-reliant one.

To be sure, influential scientists like Edward Teller, George Keyworth, James Fletcher, Fred Seitz, and Robert Jastrow tried to persuade the scientific and policy communities of the efficacy of the SDI program.[13] However, the cumulative weight of negative criticism from other parts of the scientific community (apart from the Brilliant Pebbles concept), particularly with significant press coverage of their commentaries, helped feed the mistrust in Europe about what the administration was doing with SDI. The more Machiavellian among the European commentators began to think of SDI as possibly the biggest confidence game the United States had ever tried to put over on Europe and, more important, on the Soviet Union. Just maybe, the United States was more clever than usually thought.

From Soviet public comments, especially on the margins of arms reductions talks, the Soviet Union appeared more impressed with promising SDI technologies than were the allies. General Abrahamson (director of the SDIO) briefed the Soviet delegation in Geneva on the progress of the SDI program in the late spring of 1985.

The chief scientist, Dr. O'Dean Judd, continued this practice in late 1989 and early 1990. A number of the Alliance members preferred to look at SDI as an important way for the United States to leverage the Soviets into deep reductions in nuclear arms in exchange for stopping or severely limiting the research. In short, although some of the allies thought they saw in SDI much less than meets the eye, they nonetheless hoped the Soviets saw more.

As might be expected, the coalition governments with slim majorities hinging on parties to the extremes of the political spectrum were often sparing in support of American SDI research. In private, however, the same governments proved more understanding of why the United States needed to revisit strategic defenses. The major NATO nations, especially, could appreciate the frustration of the president of the United States in the spring of 1983 over the prospects for arms reductions and the intermittent commitment, particularly in the Congress, to nuclear modernization.

NATO Political and Military Authorities. Questions about SDI raised both by political leaders and by NATO military authorities did not focus on the wisdom of research as much as on the possible development and deployment of defenses. Whether such defenses would be regional or global made little difference to the political argument. An important issue was that substantial parts of defenses would be based in space, in all likelihood including weapons. The NATO allies also looked warily on discussion of theater defenses as possibly the beginning of the space-based defense system they wished to avoid.

The argument by the military authorities had a quite different focus from that of political authorities. For example, General Rogers repeatedly complained that SDI would take away funding needed for improvements in conventional forces. Such funding in most national defense budgets was already projected to be insufficient, apart from whatever SDI might cost, to correct the critical military deficiencies that the military authorities had identified in the Conventional Defense Improvements (CDI) program in the fall of 1984.

The NATO military authorities were charged with ensuring the Alliance would have "sufficient" and "ready" conventional forces. The authorities also put some effort into areas such as sustaining war reserve stocks. Those same authorities were far less concerned about the US need to redress the vulnerability of its land-based ICBMs. After all, even if such vulnerability were a serious issue (an assessment not universally accepted), the United States was correcting the problem through modernization (MX and the new small mobile missile), passive defenses (mobility, hardening, and deception), and arms reductions negotiations in Geneva (in fact, at the Reykjavik summit the president set the elimination of all ballistic missiles as the goal). If vulnerability were the problem, then there was no rationale for SDI. After all, the Scowcroft Commission itself had earlier debunked the "window of vulnerability" in land-based systems, given the modernization programs projected and already under way.

Alliance leaders were concerned that if SDI were to get the funds requested and if it continued to enjoy presidential interest, the military-industrial drive would inevitably be to deploy "something" as soon as possible. The administration, it was feared, needed to have results from SDI to show the Congress and the public, and the pressures for deployment were building.

EXTENDED DETERRENCE VERSUS EXTENDED DEFENSE

A main difference between military and political perspectives on defenses was that some of Europe's political authorities were opposed in principle to any deployment, whether or not effective. In contrast, the military authorities did not look at defenses, even strategic defenses, as undesirable in principle. Such a predisposition would mock centuries of military thought about the nature of warfare.[14] The Alliance military authorities were also aware of how the concepts of the offense and the defense remained intertwined in Soviet doctrine.[15]

That defensive forces are inferior to offensive forces and that strategic defenses ought to be forgone for the sake of deterrence (that is, in order to remain vulnerable to the enemy) are concepts peculiar to the "atomic" age. As Bernard Brodie suggested, nuclear weapons changed the paradigms for thinking about the character of war. Every president since Eisenhower must have asked what President Reagan asked: "Is there a better way?"

Most defense commentators harbor a presumption against the feasibility of territorial ballistic missile defenses to protect populations. However, few experts would recommend in principle against ground-based point and area defenses that could protect strategic military assets—if such defenses could contribute at a reasonable cost to making retaliatory forces significantly less vulnerable. So much the better if defenses could be extended even partially to protect major population centers and the leadership of the nation or indeed of other Alliance nations. The argument, of course, is moot with boost-phase defenses since those defenses are agnostic about what targets the missiles were aimed at.

What the military authorities would eventually have to assess for political leaders would be how strategic defenses would contribute to first-strike stability, crisis stability, and arms race stability, as well as what priority defenses might have within constricting budgets. Some of the nations worried that the United States was offering defense as a substitute for extended deterrence—not a great bargain for the allies. European leaders would have none of extended defense through SDI if it would make conventional war more likely in Europe.

The military reactions to SDI differed significantly from political reactions, yet no confrontation was ever joined between the two sides of NATO's house. It was not the case that there was either appreciation or disapproval for SDI among the military but rather that no major SDI issue ever arose that needed to be resolved with unanimity at NATO. What went on within national delegations at Brussels over issues such as early deployment and interpretation of the ABM Treaty may have been a different story, but that remains unknown. Undoubtedly, the political consensus limiting SDI to research would have prevailed over any military assessment in favor of deployed defenses.

The military approach was to treat SDI as important research into defensive capabilities against a specific offensive threat, a threat that NATO had three principal means of defending against: counter-battery strikes (e.g., pre-emptive or retaliatory strikes with aircraft or missiles), passive defenses of military assets (hardening, deception, and mobility), and active defenses (point, area, regional, and even global). In other words, the reason NATO does not have any active defense against ballistic missiles, whether armed with conventional high explosives, chemicals, or nuclear warheads, is not that there is something wrong in theory with such defenses. Instead, no technologies have held much promise.

Certainly, the British worked hard to find ways to defend somehow against the German V–2 ballistic rockets that struck London from September 1944 on. The 1988 "battle of the cities" in the Iran-Iraq war showed how ballistic missiles remain a growing force in conventional wars—a force not fully appreciated. The world may well only now be in the first years of an age of proliferation of ballistic missiles—a time when a large number of nations (nearly a score by the mid-1990s) will have long-range ballistic missiles. China, for example, sold CSS–2s to Saudi Arabia, which now has the capability to strike the capitals of other Middle East countries and can even strike the Soviet homeland. Pakistan, it is reported, has ballistic missiles that may be capable of delivering high-explosive, chemical, or even nuclear strikes over long distances. The ranges of these missiles are improving rapidly and may approach 3,000 to 5,000 kilometers on some systems by the turn of the century.

That ballistic missiles can now be armed with nuclear, conventional, and chemical warheads neither changes the military strategy issue nor undercuts the potential value of defenses. In a military assessment, there is nothing inherently wrong with strategic defenses against nuclear forces of any type, except for the restrictions against territorial defenses the United States and the Soviet Union agreed to in the ABM Treaty.

NATO and the United States for decades have had defenses against some strategic threats. In the 1960s, for example, the United States had over 2,600 interceptors in the Air Defense Command to protect the US and Canadian mainland from Soviet long-range bombers attacking with either nuclear or conventional bombs and short-range missiles. The Soviets and the United States have both developed and continue to perfect means of defending against cruise missiles, whether air-, sea-, or ground-launched. Moreover, as an element in its anti-submarine warfare dimension, the maritime component of US military strategy calls for the US Navy to seek and destroy with conventional means Soviet submarines armed with sea launched ballistic missiles (SLBMs). In addition, military targets for NATO's conventional deep-strike capabilities certainly include Soviet nuclear storage sites and command and control associated with Soviet nuclear forces, as well as Soviet nuclear weapons and delivery systems.

ALLIANCE PARTICIPATION

Invitation. At the March 1985 Nuclear Planning Group (NPG) ministerial meeting in Luxembourg, then Secretary of Defense Weinberger sent a letter to NATO allies, as well as to Japan, Australia, South Korea, and Israel, inviting their participation in the SDI program. Other than the Alliance nations having to agree to the delphic language of ministerial communiques from the NPG, Defence Planning Committee (DPC), and North Atlantic Council (NAC) meetings, this was the first time nations had to do anything at home about SDI.[16]

For several nations, the invitation presented an immediate political problem, especially for coalition governments with a loose hold on longevity. Norway, for example, was quick to respond in order to get the issue tamped down without causing a split in the government. In May 1985, Norway declared that it would not participate government-to-government with the United States in the program. However, Norway would not prohibit its firms from seeking SDI contracts. After all, Norwegian officials bravely asserted, Norway was a free nation whose companies could deal as they wished in international business in the West. In the succeeding months, this formulation became the model for Denmark, The Netherlands, Canada, France, and to some extent Greece (however, Greece's position remained ambivalent, given Greek opposition to space-based defenses). Early in May 1985, President Mitterrand at the Bonn summit declared that France would not take part government-to-government in the US SDI program but would instead work on high-technology research with other European nations. At the same time, France had sufficient interest to send "teams" to the United States to look into SDI technologies and into areas France might pursue contracts in, as well as cooperative efforts industry-to-industry.

Eureka. Rather than participate in SDI research alone, a number of Alliance nations turned their attention to Eureka and welcomed France's invitation for European civilian research into high technology. The United Kingdom, Italy, and the Federal Republic of Germany decided to participate in both Eureka and SDI. By April 1985, in fact, the essentials of the Eureka concept had already been around a few years as a French-initiated collection of space and high-technology research programs, now to be dusted off and put in the spotlight. The European Community bureaucracy fervently sought to manage the Eureka projects as a competitor to overall US efforts in emerging technologies and not just those associated with the SDI program.[17] That desire continued to be frustrated as individual European nations kept the management of projects national and restricted themselves for a number of years to periodic meetings, with only a small secretariat for the program.

Some NATO governments thought they saw potential technological returns from the Eureka program, with its investigations of technologies similar to those in SDI. Best of all, Eureka did not carry SDI's political baggage. The characterization gained currency for a while that SDI was "military" (and, by implication, to be avoided), and Eureka was "civilian" (and, therefore, to be pursued). In 1985, for example, French Foreign Minister Roland Dumas characterized SDI as "a vast military program with civilian implications" and Eureka as "a vast long-range civilian program with military implications."[18] Somewhat later in the "cohabitation" government of Francois Mitterrand and Jacques Chirac, the conservative Chirac dismantled this Gallic device in favor of French participation in SDI on the basis industry-to-industry arrangements and direct contracts (US Government to French companies).[19]

The Eureka program has continued in parallel with SDI, but it has never had the concentrated management, integrated focus, and funding growth that SDI enjoyed. The United States wisely never took the French bait in criticizing Eureka in the way President Mitterrand had criticized SDI. Instead, the consistent US policy was that there was no competition, only complementaries, between SDI and Eureka.

Eureka enjoyed considerable rhetorical and some limited financial support from Alliance members. Many nations simply nominated existing national programs as part of Eureka, often without adding much funding. Unlike in the arrangements in the SDI program, participants in Eureka have had to fund the research and have committed to share the results. Moreover, the participants expect manufacturing products from the research; the work more resembles full-scale engineering development of known technologies than investigation of basic technologies. Some of the nations apparently looked to Eureka as a way to keep themselves competitive with the United States and Japan in areas such as telecommunications and computer information management systems.

Misinformation and Mistrust over Participation. In high dudgeon, political wags in Europe tried to make as much as possible out of the so-called ultimatum in the initial invitation letter. Secretary Weinberger had asked nations for a response within sixty days. In an attempt to get rid of this distracting issue, both General Abrahamson and Assistant Secretary Perle took the blame separately in public, each asserting that the mention of sixty days for a response was a mistaken judgment in drafting. In short, they meant to convey only that the program was moving along so quickly that nations needed to engage in the research process as early as possible in order to have a chance at some of the most challenging research. European complaints gave nations additional time to test the politics of participation at home, while showing their publics that they did not respond to US deadlines.

Other divisive but peripheral issues included President Mitterrand's complaint, after his mid-1985 meeting with President Reagan, that the United States was simply trying to make other nations subcontractors to US industry prime contractors.[20] US policy-makers turned this argument aside by pointing out that the US Government was the prime contractor and that therefore even American industry was a subcontractor. France and a few other nations also complained that the United States would create through SDI a brain drain from Europe. In reply, US officials retorted that the brain drain could happen only if European industries did not participate.

In general, Europeans looked at technical, financial, and business arrangements with considerable wariness. In the European view the United States was often an unreliable partner in technology transfer, in intellectual property rights, in third-country sales agreements, and in sharing advanced discoveries. Although the allies could cite many unfortunate examples to support their caution over armaments cooperation with the United States, there were also success stories such as the F–16 cooperative arrangements with four Alliance nations and the NATO AWACS program involving thirteen nations. Moreover in the SDI program, the United States was not asking for any funding from allies. In addition, because US industry was so dominant technologically in most of the areas of SDI research and because the United States was experiencing an unfavorable balance of trade with several of its European partners, some nations privately discounted ever receiving many contracts. Plans showed no set-asides for European industry; the need for competition in areas where US companies were strong led pragmatic European businessmen to keep expectations lower than those of government officials.

Agreements on Participation. By the spring of 1988, five nations (three European plus Israel and Japan) had entered formal agreements with the United States on participation. The United Kingdom was the first, on December 6, 1985. Although the Memorandum of Understanding between the United States and the United Kingdom on participation is a classified document, a list of the supposedly agreed areas of potential participation became available in early 1986—much to US consternation. Reported to be at the top of the list was "Studies for a Western European system against short-range and theatre ballistic missiles."[21] In fact, one of the early contracts the UK received was for a "theater architecture study" of what ballistic missile defenses might look like and what the system requirements would be. The UK Ministry of Defence (MOD) set up an office to coordinate the participation of government and industrial firms in Britain with the SDIO in

Washington; the Departments of Trade and Industry also coordinated closely with the MOD on every aspect of the SDI work.[22]

Arrangements for government-to-government participation were negotiated bilaterally and ultimately agreed on with the Federal Republic of Germany on March 27, 1986; with Israel on May 6, 1986; with Italy on September 19, 1986; and with Japan on July 21, 1987. As was the case earlier with the NATO AWACS program, Belgium seemed always to be two or more years behind its Alliance partners in resolving policy issues leading to participation. Belgian industries appeared eager to enter consortia and teaming arrangements for some of the work—however, the industrial associations were so inept that not much more got done. The positions of Portugal, Spain, and Luxembourg remained neutral or at least vague, with a few of the nations like Turkey, for example, stipulating that there did not appear to be any significant contribution they could make to the effort. Turkey did send a team to the United States to review the opportunities of the program. The Dutch and the Canadians also sent investigating teams to look into SDI research opportunities, but neither country was aggressive in pursuit of contracts. The desires of NATO members such as Italy and The Netherlands to forge a unified European response to the invitation evaporated as one nation after another made its own bilateral arrangements and its private decisions. Nonetheless, the Western European Union did establish a committee to study SDI issues.

The demands of the then UK minister of defense, Michael Heseltine, for a guaranteed $1.5 billion in contracts for Britain notwithstanding, the basis of cooperation with the allies remained competition by national industries.[23] Even that basis, however, came under scrutiny in 1986 and 1987 in the proposed Glenn amendment that would have put price penalties in place, favoring American industry bids over those of allies on SDI contracts.[24] If that measure had passed, the SDI agreements that waived the "Buy America" provisions of US procurement law would have come under attack. The United States would again have lost credibility in armaments cooperation.

The aggregate funding for SDI appropriated by the Congress, while substantial, remained well below the original estimated program cost of about $26 billion over the first five years, beginning with fiscal year 1985. Although the total value of contracts let to allies—late 1987 (about $125 million), April 1988 (about $250 million), and late 1989 (about $350 million)—was far less than European political leaders and industrialists wished, nonetheless that much taxpayer money going offshore at a time of severe balance of trade problems and budget deficits showed the commitment of the Reagan and Bush administrations and was a tribute to European industry for its ability to win contracts through competition.

The chart at appendix A depicts the level of participation, nation by nation, through mid-1990.

REFLECTIONS ON ISSUES RAISED
BY THE "BIG FOUR"

The United Kingdom. With her close ties to President Reagan, Prime Minister Thatcher was able to translate Alliance concerns about SDI from balance of power terms into practical considerations. She had the credibility with Reagan to direct attention to areas such as equipment procurement (read Britain's purchase of Trident), decreasing defense budgets, European elections, a sea change in the Soviet leadership with Mr. Gorbachev ("a man I can do business with"), and arms control positions. At least twice in the first five years of the SDI, Prime Minister Thatcher secured agreements, of a sudden, with President Reagan on how to proceed with SDI in the context of Soviet offensive and defensive efforts, arms reductions negotiations, and modernization of nuclear and conventional forces. That agreed language within Alliance councils became like scripture—with Prime Minister Thatcher the exegete.

Prime Minister Thatcher alone among Europe's leaders apparently had some advance knowledge of the president's strategic defense theme in Reagan's March 1983 speech. Nonetheless, that minimal courtesy, while appreciated, was no substitute for consultations on a matter of such importance to the security of the Alliance.[25] Thereafter, she took no chances that Washington would be out ahead of her on SDI; instead, she would go directly to the president every time she thought it necessary.

The first time occurred at their meeting at Camp David on December 22, 1984, a year after the completion of the Hoffman and Fletcher studies and just before the reopening of arms control negotiations with the Soviet Union. The press statement she issued after her discussion with the president carefully placed SDI within the larger context of "the prospects for arms control negotiations." Those negotiations, of course, would start with foreign ministers' discussions in January 1985, some thirteen months after the Soviets had walked out or "postponed" further participation in December of 1983.

Prime Minister Thatcher described her meeting with President Reagan, saying, "I was not surprised to discover that we see matters in very much the same light."[26] She continued, "I told the President of my firm conviction that the SDI research programme should go ahead ... we know that the Russians have already their research programme and that in the United States view that programme has in some respects already gone beyond research." This wording carefully ascribed the belief to the United States, and not to Britain, that the Soviets had gone beyond research. Nonetheless, with this as background, Mrs. Thatcher then identified four points "we agreed on."

> First, the United States and Western aim was not to achieve superiority, but to maintain balance, taking account of Soviet developments.
>
> Second, that SDI-related deployment would, in view of treaty obligations, have to be a matter for negotiations.
>
> Third, the overall aim is to enhance, and not to undermine, deterrence; and
>
> Fourth, East-West negotiations should aim to achieve security with reduced levels of offensive systems on both sides. This will be the purpose of the resumed United States/Soviet negotiations on arms control, which I warmly welcome.

The agreed language codified and protected European viewpoints: supporting, not bashing, deterrence and the strategic concept of flexible response; underlining that deployment of strategic defenses would have to be negotiated, presumably in bilateral talks with the Soviet Union, after consultations with the allies; relieving the suspicion that the United States was trying to recapture superiority through strategic defenses; and highlighting the commitment to achieve security, through arms control negotiations, at reduced levels of offensive systems.

Upon a closer look, however, the language Mrs. Thatcher worked out with the White House also had more narrow nationalistic purposes for the British government. It was not an altruistic statement on behalf of NATO members working together to solve the horrendous threat from ballistic missiles. Not to put too fine a point on the issue, a main contribution of the language was to protect the British decision to proceed, at a cost of about $13 billion, with the purchase of the Trident submarines and associated ballistic missiles (SLBMs).

Mrs. Thatcher and her ministers were often less than comfortable during question time in the Parliament or during national security debates with the likes of Neil Kinnock, Denis Healey, David Owen, and David Steele to have President Reagan quoted as calling into question morally and strategically NATO's current deterrence policies. Moreover, President Reagan continued to favor the elimination of all nuclear weapons or at least the attempt to render ballistic missiles "impotent and obsolete." His positions seemed more akin to those of the Greenham Common women protesting GLCM deployment than those of the tough-minded leader of the Alliance in Europe, namely Mrs. Thatcher herself.

In Mrs. Thatcher's view, Soviet achievement of increased capabilities to defend its territory against ballistic missiles over the next few decades would undercut the effectiveness of the independent nuclear forces of both Britain and France. She had to ensure that such defenses would not move to the top of the Soviet agenda as a result of SDI and antisatellite (ASAT) work in the United States, and that US desires to eliminate ballistic missiles would not undermine her rationale at home for procuring such missiles. The costs for Trident submarines and D–5 missiles, as the Labour party never ceased to remind her, were making the British defense budget creak. The thought of adding countermeasures to increase the survivability of the warheads and to ensure their ability to reach targets in the Soviet homeland was too much to contemplate.

The president, even in his initial SDI speech, recognized that offensive nuclear forces would remain the main underpinning of deterrence for a long time to come. What he envisioned was a gradual transition to strategic defenses, negotiated deep reductions in the offensive arsenals, and modernization of the offensive systems themselves, especially command, control, and communications. Amid the barbs and jibes of question time, this approach was not nearly as effective as to say that research would continue, but that the president had agreed that nothing would happen for a long time to come. Furthermore, the allies will be involved at every step of the way. This point was the real achievement Mrs. Thatcher had garnered in what she agreed on with President Reagan, even though Washington may not have thought she gained anything significant. With these points in Mrs. Thatcher's grasp, her foreign secretary could ask tough questions about SDI deployment, feasibility, and affordability.

In a March 1985 speech, Howe posed difficult questions that struck at the core of SDI assumptions. However, his questions had little effect on Britain's own narrow decision nine months later to participate in the research and to expect (thank you, Mr. Heseltine), a large share of the contracts for the privilege. In Britain itself, the Labour party opposed SDI, as indeed did most opposition parties in European parliaments.[27]

At the same time, however, SDI did enjoy editorial support from the *Times* (London) and the *Economist*[28] as well as from influential conservative spokesmen such as Lord Chalfont and retired senior military officers such as Air Vice Marshall Stuart B. Menaul (former chief of staff of the R.A.F. Bomber Command).[29] The *Times* in particular took the attack, for instance, against Foreign Secretary Geoffrey Howe, calling his speech "mealymouthed, muddled in conception, negative, Luddite, ill-informed and, in effect if not intention, a 'wrecking amendment' to the whole plan."[30]

Chalfont, Menaul, and a few others sought to remind Britain of its stake in research and even deployment of defenses against ballistic missiles, whether armed with nuclear, chemical, or high-explosive conventional warheads. Menaul, for example, was one of the originators of the concept of a "European Defense Initiative" for theater defenses to complement SDI-derived global systems. According to Menaul, the concept of trying to defend against ballistic missiles "really began not in 1983 with President Reagan's momentous statement, but on [September 8th,] 1944, when the first ballistic missile fell on London."[31] Over four thousand V–2s fell in the area of London—some eight million pounds of TNT. Menaul explained, "Throughout history you have had in almost every case some form of balance between offensive and defensive weapons except for the ballistic missile which has had its own way since 1944."[32]

While Alliance members have had no difficulty in developing and deploying extensive defenses—active, passive, and counter-battery—such as air defenses against airplanes and cruise missiles (whether or not they are nuclear-armed)—there continues to be great reluctance even to think about active defenses against strategic and theater ballistic missiles. Nothing in in the nature of the ballistic missile itself—not its promptness, multiple warheads, accuracy, range, and destructive potential—fully explains this curious attitude toward defenses. Rather, the explanation needs to be found in the externalities that provide the context within which ballistic missiles contribute to deterrence. It is almost as though the psychology of strategic deterrence, in light of the horror of nuclear war, precludes any worth (in the West at least) in even thinking about defenses.

One constant in the pleas of British critics and supporters alike, both inside and outside of government, was for "predictability" in US policy shifts on SDI. The US record was spotty at best from Whitehall's perspective: from the initial announcement by the president, through the loose discussion of the broad interpretation of the ABM Treaty in the fall of 1985, and to the discussion in the United States of early deployment in 1986. In each case, the consultation process often left much to the imagination; however, the governments at senior levels, received many more SDI updates than they acknowledged in their complaints about consultations.

After early 1985, on routine SDI matters, the United States received high marks for consultations with the allies, particularly through the efforts of Ambassador Abshire, Ambassador Nitze, the US negotiators from Geneva, Assistant Secretary Perle, and Lieutenant General Abrahamson (Director, SDIO). However, whenever it came to new directions in SDI, there was little or no exchange of views with the allies in advance. What discussion there was often had the character of "hasty pudding" explanation, not consultation. The good will built up in the interims was lost each time in the mini-crises of summitry and in Alliance divisions that the international press members were quick to exploit.

One example of the failure of communications was the way, from an Allied viewpoint, that things got out of hand during exchanges between President Reagan and General Secretary Gorbachev at the Reykjavik summit of November 1986. Because of the many issues of Reykjavik,[33] Mrs. Thatcher quickly booked passage to Washington after the summit to put order back into the agreed NATO approach to arms control and to SDI, as well as to lessen the confusion and soothe the antipathies building in Europe over the strange and contradictory accounts of what did or did not happen in Iceland. In particular, even the suggestion that the president wanted to be rid of all ballistic missiles—or perhaps all nuclear weapons—within ten years caused the winds of unsettling change to rage across Europe. Especially chilled were the spines of political leaders in Great Britain and in France.

For months after Mrs. Thatcher's trip to Washington, British officials, such as Sir Geoffrey Howe and George Younger (then secretary of state for defence), repeated the "agreement" between the president and the British prime minister, reached at Camp David in late November 1986. The main points of that agreement served to put the brakes on the potential and momentum of arms control positions in NATO to move toward the denuclearization of Europe. Despite the talk of a world without nuclear weapons at Reykjavik, Mrs. Thatcher got Mr. Reagan to agree to the critical importance of offensive nuclear weapons for deterrence.

Along with other points, the agreement included the commitment that "SDI research should continue within the terms of the Anti-Ballistic Missile Treaty." Moreover, "these matters should be the subject of continued close consultation within the Alliance."[34] These same concepts, with some variations, showed up in the Euro-Group ministerial communique on December 3, 1986, and in the DPC and NAC communiques a bit later in the same month.

France. It may be tempting for some commentators to dismiss French reactions to SDI as typical Gaullist spurning of any policy not directed from, if not in fact conceived in, France. However, French policy views are not easily dismissed. They usually give careful listeners not only fits but also pause because there is often more substance than at first may appear. Troubling issues did underpin the nettlesome criticism from Paris. The French critique of SDI began early and remained negative, despite the softening of some of the positions on participation in the research within the cohabitation government with Mr. Chirac. Like that of several other nations, French questioning focused on the feasibility and efficacy of space-based defenses, worry about isolationist impulses in America, the expense of SDI deployments that would draw off funding needed for conventional defense improvements, concerns about limited chances if SDI were deployed for reductions in the strategic nuclear arsenals of the superpowers, the increased likelihood of conventional conflict in Europe, and the probability of an arms race in defensive systems as the Soviets attempted to keep up with the US efforts.[35]

What was curious about the criticism was France's recidivist support for NATO's strategic concept of flexible response—the same McNamara strategy that General Charles de Gaulle had rejected in the early 1960s.[36] Assuming eventual Soviet achievement of nuclear parity with the United States, de Gaulle had scorned the notion of extended deterrence. He found it an absurd concept that the United States would risk the sovereignty and survival of its homeland for its allies by attacking the Soviet homeland with massive nuclear strikes in retaliation for Soviet strikes limited to Western Europe. In de Gaulle's view, this logic made no sense to a France and to a Europe that cut its teeth during an era in which the nation-states continually shifted allegiances to create favorable balances of power. No nation would risk its survival for another, no matter how strong the alliance.

Once the Soviet Union began to acquire the capability to attack US targets with intercontinental ballistic missiles, de Gaulle determined that France would have to rely primarily on itself for nuclear "dissuasion" of Soviet attack. To this end, he created a small *force de frappe* of nuclear weapons and delivery systems strong enough to damage the Soviet Union severely (probably by attacking targets such as industries, leadership, and cities) in retaliation for attack on French soil. Although French declaratory policy remains ambivalent, France has implied that its nuclear response would be not solely to Soviet nuclear attack but also to any direct attack on the sovereignty and survival of France. The French nuclear force, currently undergoing a three-fold increase in warheads in its modernization program, relies heavily for its effect on the ability of substantial numbers of its ballistic missiles to strike their targets. If the Soviets were massively to increase their efforts to defend their territory against ballistic missiles as a result of the US SDI program, then the consequences could eventually be devastating to the credibility of the *force de frappe*.

The oddity of France as the defender of the orthodoxy of NATO's flexible response was not lost on the other members of the Alliance. Some wondered whether there was not just a bit too much convenience in France's use of the argument that the superpowers should remain vulnerable to ballistic missile attack. Echoing through Europe in this debate were the voices of Chancellor Adenauer, President Eisenhower, Prime Minister MacMillan, and Le Grand Charles—rehearsing the old verities of nuclear weaponry and concepts of "dissuasion" and deterrence, first in the 1950s, as well as in the early 1960s with President Kennedy and Secretary McNamara.[37]

It had taken McNamara from 1962 to 1967 to persuade the allies of the importance of replacing "massive retaliation" with the strategic concept of flexible response, including deliberate and controlled escalation. By 1967, however, McNamara himself probably no longer believed in the possibility of controlling escalation. "Assured destruction" (namely central strategic systems) was all that McNamara could contemplate as the basis for deterrence. In his view, the ABM Treaty of 1972 codified this understanding even though he had had such a difficult time explaining "vulnerability" to Kosygin at Glassboro in 1967. The French government never acknowledged the McNamara theory of deterrence, and senior German and French leaders today, not just those out of office, fully understand the flaws in the Alliance's nuclear strategy—despite agreement to the General Political Guidelines in 1986.[38]

Former Chancellor of West Germany Helmut Schmidt, for example, stated that it is "a ridiculous illusion" to believe that the "German Bundeswehr will still fight on after you have eradicated Nuremberg or Frankfurt or some other city" in Germany with nuclear weapons. He continued, "Neither the NATO generals, nor the French, have asked themselves enough" questions about who would fight on after even "limited" nuclear attacks by the West in the West or by the East in the West.[39] McNamara has admitted that he had advised Presidents Kennedy and Johnson never to use nuclear weapons first, undercutting a fundamental option in the strategy of flexible response. In Schmidt's words, "I did not find ... and I have never seen since, any indication that anyone in the world knows how to initiate the use of nuclear weapons with advantage to NATO."[40]

The notion of a United States protected against attack from Soviet ballistic missiles while Europe remained unprotected deeply disturbed French officials almost from mid-1984 when SDI criticism began in Europe in earnest. While de Gaulle could pull out of the integrated military structure in 1966 and establish the small but independent nuclear *force de frappe*, he could also count on the residual effects, if any, of extended deterrence from the US central strategic systems. The systems, after all, held at risk targets in the Soviet homeland. With SDI deployed and with similar Soviet strategic defenses, France no longer could have it both ways. In August 1984, the French foreign minister, Claude Cheysson, compared the SDI concept of space-based protection to the Maginot Line.[41]

The then assistant secretary of defense for international security policy, Richard Perle, tried on several occasions to turn the French argument aside by asserting that "a defended America is more likely to be understood by the Soviet Union to mean it when it says it will defend Europe with nuclear weapons." However, Perle's counterargument was "small beer" to Europeans who saw SDI protection as an isolationist desire on the part of the Americans—not, in fact, generically different from calls for massive withdrawals of US forces from Europe (e.g., Senator Nunn's June 1984 amendment). However, more clever was Perle's retort that SDI conceptually was a layered defense in depth established far enough forward to protect the territory of the allies as well as that of the United States. "A better way to look at it is that we are attempting to put a dome over the Soviet Union that will keep Soviet missiles in."[42]

The strategy issues provided the backdrop for several second order French arguments against SDI—the arguments (discussed elsewhere in this text) about the importance of the civilian Eureka program, the potential brain drain to the United States, and the concern that the United States would outperform Europe for decades from leaps forward in certain technologies. Certainly a more important issue was modernization of French nuclear forces. Only rarely, however, did French officials object that potential Soviet strategic defenses against ballistic missiles, spurred by SDI efforts, would undercut the French nuclear deterrent and render futile France's modernization program.[43]

Federal Republic of Germany. Given the French and British positions, the Germans were in the unenviable state of recognizing the importance the US administration attached to SDI but needing to support the views of France and Britain. The French would not entertain even the possibility of a change in "dissuasion," based as it is on the ability of French nuclear forces to strike the Soviet homeland. No matter what contribution strategic defenses might make, the offensive nuclear arsenal and the willingness to use that force *in extremis* were what created the psychological state to dissuade Soviet leaders. The British were also adamant that there be no premature discussions of a new strategy. A research program without known results seemed a particularly poor rationale for the United States to use in justification for a fundamental review of deterrence. At the same time, the British were characteristically more willing than the French to try to guide the United States into thinking that the British approaches were made in America.

The Germans could not afford to appear any less skeptical about SDI than the French and the British. Like the United Kingdom, therefore,

Germany in the spring of 1985 joined the French-inspired Eureka program. Germany was no more sympathetic to SDI than France and the United Kingdom but on occasion appeared to be.[44]

German agreement to government-to-government memoranda of understanding on SDI participation came on March 27, 1986—some three months after UK accession. Influential industrialists in Germany, representing major companies and consortia, in some cases already engaged in long-term teaming arrangements with US companies, supported participation in the SDI from the late spring of 1985. These pressures no doubt influenced German political leadership. For example, a number of industrial conferences on SDI took place in Germany, sponsored by US think tanks such as the Institute for Foreign Policy Analysis and by German organizations such as the Konrad-Adenauer-Stiftung.[45]

German support for participation, of course, did not extend to deployment of global, space-based defenses against strategic ballistic missiles. Like other NATO nations, German approval was pointedly limited to research within the traditional interpretation of the ABM Treaty. In the coalition government of Chancellor Helmut Kohl, there continued to be less than enthusiastic support for SDI by his foreign minister, Hans Deitrich Genscher, from the Free Democratic party. With deep personal and political interest in Ostpolitik, the arms control process, and a period of "new detente," Genscher was forever fretting about the war-fighting potential of strategic defenses. He was also unconsoled by thoughts of the potential effects of strategic defenses on Soviet decisions to push ahead with their own defenses and to deploy offensive countermeasures such as increased numbers of warheads. Beyond this foot-dragging in the German executive, there was also outright opposition to SDI in the Bundestag by members of the Social Democratic party (SPD) of former Chancellor Schmidt.[46]

The political juggling found its counterpart in scientific debates about the efficacy of SDI technologies. Like the exchanges in the United States between Hans Bethe and Edward Teller and those between Richard Garvin of the Union of Concerned Scientists and SDI proponents such as General Abrahamson, Robert Jastrow, and James Fletcher, in Germany Hans Ruehle (head of the planning division of the Ministry of Defense) squared off in the popular magazine *Der Spiegel* in 1985 against Hans-Peter Duerr (director of the Max-Planck-Institute for Physics and Astrophysics), who had written the "credo" of German scientists against SDI.

Without belaboring the debate, suffice it to say that Kohl and Woerner, through the voices of Ruehle and others, insisted even in early 1985 that SDI get a fair hearing in Germany, no matter who the opponents were. The most serious "error" in the Duerr approach, according to Ruehle, was to assume that "defense systems must be perfect to guarantee, or at least improve, security and strategic stability."[47] In other words, even limited strategic defenses of military assets can enhance deterrence.

In his *Star Wars: Strategic Defense Initiative Debates in the Congress*, Senator Pressler accurately described the pressures Kohl was under politically to keep Germany at the table of high technology with both European and American partners and at the same time to watch carefully the pace of French and British reactions to the progress of the SDI program.[48] In negotiating the agreement on participation in the program, for example, Kohl chose to put as "civilian" a face as possible on German involvement: keeping the government out of direct association as much as possible, selecting as the lead agency the Economics Ministry (a Free Democratic party—led by Dr. Bangemann of Minister Genscher's party) and not the Defense Ministry (then led by Minister Woerner), and generally playing down expectations, in contrast to Michael Heseltine, who tried to increase expectations of contracts in England. Kohl's political instincts led him to follow the British in circumscribing political support for SDI. The German team was dominated by Horst Telschik, the chancellor's right-hand man.

Chancellor Kohl, like Prime Minister Thatcher, periodically established a package of "points" about SDI and about German participation. Whenever a new policy issue was raised, the Kohl government would return to the set of bedrock guidelines it had set for itself as a way through the labyrinth involving sensitivities of the United States, the major Alliance partners, NATO itself, and the Soviets as well as the coalition government in Germany that did not want SDI to become a major domestic political issue.[49]

While there were variations and different emphases to Kohl's points from early 1985 on, several themes paralleled French and British concerns. Perhaps the most important of the guidelines was that NATO's current deterrent strategy based ultimately on central strategic systems was valid and would remain so for a long time. President Reagan, of course, had challenged strategists to think about the possibility of deterrence based more and more on defensive systems. The president recognized that it might take decades to come up with technologies that would allow a change in strategy. Therefore, despite the enthusiasm some SDI zealots might exude about how fast SDI was progressing, the administration could, without being disingenuous, second Kohl's insistence that SDI was the investigation of only a possible alternative to the current strategy.

In fact, an early 1985 White House pamphlet on SDI endorsed the idea that NATO's strategy would continue to be based on offensive nuclear forces for a long time to come.[50] Predating Kohl's address on SDI to the Bundestag in April 1985, the pamphlet reaffirmed that "NATO's strategy of flexible response, which is the basis for deterrence and peace in Europe, remains as valid today as when it was first adopted in 1967." [51] As an aside, this assertion, of course, begs the question of the strategy's validity in 1967.[52]

Europeans know where the shortcomings of the strategy have been papered over for political purposes. They also recognize that this strategic concept, despite its origins in Ministerial Guidance and in NATO's Military Committee (MC 14/3), is a political creature, not a military strategy, in the strictest definition. It is a strategy of accommodations. Every government understands that if war did begin, new military arrangements in pursuit of wartime political objectives would quickly evolve through consultations to meet the unfolding attack—whether or not forward defense were maintained and whether or not there would be early use of selective nuclear options. The NATO strategy remains a peacetime strategy for deterrence; it also spells out the desired objectives for fighting and terminating war if war were to come. In this tension between deterrence and war plans, SDI defenses can at once find blame and acclaim.

In an interview with Elizabeth Pond of the *Christian Science Monitor* in April 1986, then Minister of Defense Manfred Woerner captured some of the tensions between SDI defenses and deterrence. In warning the West not to trash its own strategy prematurely, Woerner said he was fighting "one danger" in particular, the danger that "we destroy intellectually a strategy which has no alternative and which will prove to be efficient for at least fifteen years from now. We have to reflect on the possible consequences—but we need to know that these are consequences in the future and not tomorrow. So I believe in the efficiency of nuclear deterrence."[53]

The second point that Kohl, as well as Woerner, often made was that there must not be "zones of different security" in the Alliance. That is, if only the territory of the United States and perhaps that of Canada were protected through SDI from ballistic missile attack, there would not be the same vulnerability or protection across the Alliance territory. To that extent, the United States might feel less "coupled" or less committed to the security of Europe, a Europe that would provide the battlefield both for conventional war and for nuclear war. This vulnerability issue helps explain German interest in anti-tactical ballistic missile (ATBM) efforts. Within Germany, however, there was also considerable opposition to ATBM for strategy, financial, and arms control reasons.[54]

Kohl's third guideline elaborated a thought implicit in the British points; namely, the Alliance must avoid instability in the strategic balance with the Soviet Union during any transition to any new relationship more dependent on defensive systems. One of the Soviet and European complaints was that the SDI would cause an arms race. The Soviets would match US efforts in defensive systems, increase the number of ballistic missile warheads, and take other countermeasures. In contrast, the United States contended that the Soviets were well along with comprehensive civil defenses, an operational anti-satellite (ASAT) capability, air defenses effective against some tactical ballistic missiles, a comprehensive modernization of the ABM systems protecting Moscow, and extensive research into the same strategic defense technologies that the United States was investigating. Since these positions were inconsistent, the allies ensured that the United States understood they would not support efforts that would create dangerous periods of instability during any transitions.

Kohl insisted that the deployment of defenses would be of small value to the Alliance if threats below the level of nuclear forces were to increase as a result. Even among those who knew better, the strange notion grew in NATO that a deployed SDI would also mean the elimination or at least the diminution in importance of *all* nuclear weapons in the theater, leaving the allies to face the brutal reality of the Red Army reinforced by its Warsaw Pact allies (at least as long as the Soviets were winning). This approach (a constant until the 1989 revolutions in Eastern Europe) ignored the presence of non-ballistic nuclear weapons in the theater, as well as in the arsenals of France, Britain, and both superpowers.

In addition, Germany, too, assumed that it was better to live with the threat of annihilation than to ensure that conventional forces and non-ballistic nuclear forces remained strong enough to deter once defenses were deployed against ballistic missiles. Little attention was given, even in the Follow-on Forces Attack (FOFA) context, to the potential contributions of SDI research and SDI-derived systems to conventional defense—battle management, surveillance, and reconnaissance. Not too far out of Kohl's consciousness, no doubt, was an appreciation for the intentions of the French and the British leadership to protect their independent nuclear arsenals. Kohl did not want to be excluded from British, American, and French discussions of nuclear cooperation the way Adenauer had been by de Gaulle.

In the background of European reactions, especially the German response, was considerable pique that the United States did not appreciate the political difficulties of the nations deploying the intermediate-range nuclear forces. The SDI came in the "Year of the Missile"—for some the "Year of the Anti-Missile." Yet the United States was expressing doubts about the credibility and morality of nuclear deterrence, just as European governments were beginning to deploy the intermediate-range ground-launched cruise missiles (GLCMs) and the Pershing IIs. In the foreword to the White House pamphlet of January 1985, the president said, "We must seek another means of deterring war. It is both militarily and *morally* necessary." This rationale still prevailed in the fall of 1986 in the White House "Issue Brief on the Reykjavik Summit," asserting, "SDI offers a safer and *more moral alternative* to deterrence of nuclear attack through the threat of retaliation."[55]

Despite Ambassador Nitze's effort to handle the issue, this high-minded rationale lingered in the explication for SDI (not that any NATO government wished to deny the proposition although some may have expressed it differently). The NATO members simply felt that nothing should undermine the current strategy unless a better strategy became available. That certainly was not the case with a research program.

Like the British, who insisted on predictability from the United States in regard to SDI policies, the Germans tried also to caution against "surprises." The Kohl government was willing to work with Washington on SDI (that is, without any financial contributions) but did not want to be faced with sudden announcements about controversial SDI issues. At the same time, Germany continued through the 1980s to hedge its bets with participation in Eureka programs, bilateral defense talks with the French (including the proposed establishment of a joint brigade), the "new detente" with the arrival of Gorbachev, endorsement of the concept of a "European pillar" that German leaders in and out of office had been touting, and a wistful glance now and then in the direction of wished-for extended deterrence by the French.

Italy. As with other major NATO issues, Italy warily watched until the moment seemed right for her to engage SDI politically. Italian industries and research institutes were technologically capable and wanted to participate in projects. They submitted more than seventy unsolicited proposals and responses to requests for proposals (RFPs) by the spring of 1987. Italian government leaders preferred to follow the lead of France, Germany, and the United Kingdom. Clearly, significant contributions to SDI research would be within the competencies of those nations. If they chose not to participate while Italy did, then Italy would have to stand out from its partners; however, Italy prefers company in most Alliance commitments. In contrast to the gingerly steps of government leaders, Mr. Agnelli, chairman of the board of FIAT, was an early and ardent advocate of SDI.

Italy was in the forefront of nations wanting to establish a single European response to the United States, strongly preferring the newly revived Western European Union (WEU) as the right forum to generate a common position. Nothing of substance ever happened. After all, for the WEU to establish such a voice would mean a diminution of sovereignty added to the warrants on sovereignty granted earlier to the Common Market, the European Parliament, and NATO. After the British and German accession, Italy moved ahead with negotiations and joined the program in September 1986.

Italy shared many of the same apprehensions her partners expressed about the implications of SDI. However, several reactions made Italy's responses different from those of other major NATO nations. First, although Italy occasionally would chant the ritualistic European list of objections to deployed strategic defenses, in Italy's view there was no doubt that potential economic gains and technological spin-offs from the research were benefits that Italy should participate in. After Italy's initial decision to participate, SDI was never a recurring issue in the Italian Parliament. There was never much worry in Italian commentary that SDI would ever progress beyond research—and certainly not within the time allotted to the Reagan administration. Italy had no political difficulty in strongly supporting the research efforts as a hedge and as insurance, the principal US rationales after early 1985 to justify the technology investigations.

Moreover, there was almost a bantering elbow-in-the-ribs quality to Italian views of the potential of SDI to exact a high price in deep Soviet reductions in offensive strategic nuclear systems at the negotiations in Geneva. In private, at least, there were almost theatrical winks from interlocutors whenever US spokesmen stressed that the administration would never treat the SDI as a bargaining chip. As did most of the allies, Italy accepted the conventional wisdom that it was SDI that had driven the Soviets back to the negotiating table after they had walked out in the fall of 1983.

The Italians appeared to believe that the Soviets were taking the possibility of deployed "thoroughly reliable" SDI systems more seriously than the allies were and that the Alliance should do nothing to disabuse the Kremlin of that perception. With this viewpoint, Italy was more concerned about the eventual effects on arms control, specifically the ABM Treaty, than it was about the implications for NATO's strategy.

Italy expressed considerable interest in the possible spin-offs of the SDI technologies to commercial applications and to improvements in conventional military capabilities, from foxhole trencher to satellite. Perhaps more than any other NATO member, Italy saw nearly limitless, worthwhile potential in the technologies and appreciated the work needed to make that research successful. Ever the close ally, Italy did not attribute sinister motives to the United States in pursuing SDI, gave the United States the benefit of the doubt that consultations would be adequate, accepted US assurances that the SDI had been structured to stay within the ABM Treaty, and sought to study ways in which SDI might help with deterrence and defense in the European theater.

Particularly before the INF Treaty, a number of nations thought that SDI technologies might be translated into anti-tactical ballistic missile defenses. Investigating a purchase of Patriot at the time—an updated Patriot with some ATBM capability—the Italian government was cautious lest SDI be used against a Patriot buy, betting on the arrival of a system useful against planes and tactical ballistic missiles. In this context, Italy worked closely with Germany, within extended air defense efforts, to understand the projected threat from and effective defenses against Soviet tactical ballistic missiles. Moreover, under contracts let within the US Army Strategic Defense Command's efforts—paid for by the SDIO—Italian companies entered teaming arrangements with consortia studying theater architectures and investigating battle management for missile defenses. For obvious reasons, Italian efforts concentrated on the ballistic missile threat to the Southern Region.

On September 19, 1986, Italy reached agreement with the United States on participation in the SDI program. As with Germany, Italy had its economics and foreign affairs ministries involved in the discussions and in the follow-up machinery to win SDI contracts. Some Italian firms were so confident of the research that they, almost alone among the European companies participating, contributed their own funding to SDI research projects.

Unfortunately, Italy has not won anywhere near the number and quality of contracts that it had anticipated. To some extent, the causes for this frustration lay at Italy's own door. Although Italy forwarded a large number of research proposals, they were either not timely or they concerned areas already covered or they simply did not fulfill the requirements of US law and regulations. By the spring of 1987, for example, Italy had received only about $2 million worth of contracts from the SDIO. Italy grumbled that this was not much of a return for the political capital it had expended in supporting the US efforts. One of Italy's own failings was in not having in Washington the right people to help keep Italian firms responsive to requests for proposals. By early 1988, Italy was doing a much better job in an area so long the province of those from the United Kingdom, France, and Israel.

ISSUES IN RESERVE

The Alliance capitals returned to a few important themes from time to time, almost like territorial markers. For example, the Benelux nations, along with the Nordic nations, strongly rejected even the possibility that arms might be deployed in space. They believed that this was the one place that remained empty of armaments and that it should remain so. They understood that the "militarization" of space had begun as early as 1957–58 with *Sputnik* and with the first flight of an ICBM, if not in fact with the first flight of a V–2 in 1944. Moreover, they also recognized that both the Soviet Union and the

United States used space for military purposes such as surveillance, communications, command and control, verification of arms control agreements, weather monitoring, and navigation. However, the fact of arms permanently in space was quite another matter.

Alliance nations also never missed a chance to remind the United States not to expect any help with funding. In their view, through mid-1989, all available defense funding in Europe should go toward improving conventional forces. If there would be any funding available for defense against ballistic missiles, it should be applied to terminal defenses in the theater (for example, Roland and Patriot follow-ons). There is now in 1990 little question of where German funding will be spent.

A number of nations also tried to ensure that the United States kept its commitment to the Alliance, not to seek superiority with SDI defenses. The allies believed the focus in modernizing offensive and defensive strategic forces should be on enhancing the survivability of a second-strike capability. To this end, the superpowers might agree on a stable transition in which each would gradually deploy defenses as they both reduced the numbers of offensive forces, particularly those with first-strike potential. For all the nations—not just Denmark and Greece with their strong views on arms in space but also the United States and Canada—whose territories are strategic targets, "stability" in any transition was the paramount necessity.

From strategic modernization to arms control, from strategy issues to potential transitions, and from a deterrence based on assured destruction to a deterrence based more on defensive systems for mutual assured survival, there were no easy issues with SDI. The president's challenge singled out the core concepts of nuclear deterrence and ultimately of the Atlantic bargain itself—deterrence versus defense. We need now to turn to the implications of SDI for both.

4. IMPLICATIONS FOR NATO'S STRATEGY

From Gorbachev's standpoint [by the December 1987 Washington summit], Reagan's attachment to SDI had become less a threat perpetuated by a dangerous adversary and more an object of indulgence, the fanciful obsession of an eccentric lame-duck President whom Gorbachev could afford to humor.

—Strobe Talbott

In the months that followed the October 1986 summit at Reykjavik, Iceland, between President Reagan and General Secretary Gorbachev (a summit Strobe Talbott characterized as "one of the most bizarre encounters in the history of diplomacy"), the Soviet Union began to adopt a much less petulant stance towards SDI.[1] Probably by the December 1987 summit in Washington and certainly by the June 1988 summit in Moscow, the Soviet Union and the NATO allies understood that nothing radical would happen during the Reagan administration regarding SDI decisions on development and deployment of defenses. Either signatory could have raised contentious issues at the regular five-year review of the ABM Treaty, and some in the Reagan administration had wanted to give notice of US withdrawal because of the clear Soviet violation of the treaty with construction of the Krasnoyarsk radar. However, by the June summit, it was clear to the Soviets that US talk of termination could be discounted in the last months of Reagan's presidency. The outcry from those seeing abrogation as irresponsible, including the European allies, would have been too much to bear so close before the US elections.[2]

The US Congress substantially increased appropriations for SDI from fiscal year 85 to fiscal year 89. At the same time, however, Congress had not funded it anywhere near the level needed for early deployment of even thin terminal defenses, let alone space-based sensors, kinetic interceptors, and battle management for a global system against ballistic missile attack. Moreover, led by Senator Nunn, the Congress by mid-1987 had prevailed over hardliners in the administration in regard to the "narrow" versus the "broad" interpretation of the ABM Treaty. There would be no development, no testing, and no deployment of technologies based on "other physical principles" (under Agreed Statement D of the ABM Treaty) beyond those SDI projects already planned to be conducted in compliance with the so-called narrow or traditional interpretation of the treaty. These prohibitions sponsored by Senator Nunn technically applied only to fiscal year 1988—at least at first. But there simply was no strong political consensus anywhere to go beyond the traditional interpretation.

With the arms control climate much improved through ratification of the INF Treaty, with favorable possibilities for START, with only months of the Reagan presidency remaining, and with major presidential candidates promising either to change SDI radically or at least to slow down the research, there was indeed less for the allies and for the Soviets to worry about from mid-1988 on. In fact, for advocates of ballistic missile defenses, the problems by the spring of 1988 were how to salvage SDI projects and how to take advantage of the residual political support for SDI among the "people" in the United States in order to get even a weak commitment to some deployment.

To be sure, the political situation—not only in domestic US politics but also in East-West relations paced by a multitude of Gorbachev's initiatives and the revolutions in Eastern Europe—grew decidedly less tense between mid-1983 and mid-1990. Nonetheless, the strategy issues raised by SDI would in time still have to be dealt with, perhaps with less anxiety than when a few NATO nations had earlier called for a formal study of SDI's implications. Moreover, the question of transitions to a strategy more reliant on defensive capabilities would not disappear from the agenda even if the priority of SDI were much diminished.

The "defense" factor is a constant in US politics and policy. In fact, the anti-nuclear motif of the tough Republican president in some of his administration's justifications for SDI was not generically different from that of the so-called weak Democratic President Jimmy Carter. For example, in his inaugural address of 1977, President Carter prophesied, "we will move this year a step toward our ultimate goal—the elimination of all nuclear weapons from this earth." Perhaps administrations show more continuity in US foreign policy and in the military strategy and arms control components of that policy than is often recognized in the friction of political contests. Whether rhetorical flourish or steadfast principle, the topic of elimination of nuclear weapons has many standard-bearers in both major political parties—from the right and left of the policy spectrum.

NATO'S STRATEGIC CONCEPT

The question of SDI's implications for NATO's strategy relates more to debate within the Alliance than it does to the strategic relationship between the United States and the Soviet Union. The sometimes overlooked point is that the concept of flexible response has been NATO's and not the Soviet Union's. In a 1985 interview with the BBC, General William Odom, then director of the National Security Agency, argued that there was no analogue for the Western concept of deterrence in Soviet military writing, saying, "There are so many paradoxes (philosophical, moral, military, and others) in deterrence theory, it's amazing that this paradigm has succeeded as long as it has."[3] The European argument that SDI would undermine flexible response by reducing the credibility and changing the character of limited nuclear options would make no sense to the Soviets in the first place. As Odom continued, the deterrence strategy of NATO would "strike a Soviet, a serious Marxist-Leninist, as a simple-minded subjectivism, or bourgeois idealism, whereas an objective war-fighting capability that has some campaign success, even at great losses to your own society, is an objective capability, not simply a subjective capability."[4]

Former Secretary of Defense Robert S. McNamara has spoken at length in the last few years about the origins and meaning of flexible response. McNamara's version of the strategy, as originally worked out by General Maxwell Taylor and others in the late 1950s, bears resemblance only in outline to the allies' understanding of the NATO concept delineated in the Military Committee document (MC 14/3).

Implications for NATO's Strategy

The strategic concept that the allies approved in 1967 and later adapted for their own ends does not match up verse for verse with what McNamara now believes were the motifs of the strategy. In his own words, he now asserts that the United States in the mid-1960s was "seeking to move away from 'massive retaliation,' to replace it (a) with conventional response to conventional attack, and (b) to the extent that nuclear weapons were to be used, to use them late and in limited quantity, and against military as opposed to population targets, in order to limit the Soviet nuclear response and thereby limit the damage to NATO."[5] The concept was not that the homeland of the United States should be exposed in the first instance for the security of Europe, but that the US homeland would be at risk for the security of the United States and only by extension for Europe's security. The United States was at risk not because any strategy required it and not because any strategist wanted it that way but because there was no effective way to preclude that vulnerability, given developments in the Soviet nuclear arsenal.

This difference in recognition is at the heart of the debate over SDI in Europe. In McNamara's understanding, the intent and hope of flexible response were not only to terminate the conflict but also to limit the *Soviet response* and to limit *damage* to NATO (read Europe) in the event of war. Geography itself dictated different zones of security within the Alliance, with Europe obviously vulnerable to conventional war and vulnerable to attack with theater nuclear weapons in ways that the United States and Canada were not.

Flexible response was thought of in the United States as a sub-strategy for the European theater.

The global strategy in McNamara's mind was not "flexible response" but "assured destruction" whereby both the Soviet and the American homelands and populations were vulnerable to destruction so devastating as to preclude the use of nuclear weapons in the first place—in whatever theater. McNamara's scenario was that the Soviets would strike first and that it made little sense in the 1960s to aim at counterforce land-based missile targets in retaliation—i.e., empty silos. Because any nuclear use in the European region could escalate to a central exchange, the theory supposed that deterrence would obtain in the theater as well. This was the "extension" of deterrence McNamara had in mind. In the 1980s, along with the other members of the "Gang of Four" (McGeorge Bundy, George Kennan, and Gerard Smith), McNamara struggled against NATO's reliance on the early use of nuclear weapons and strongly advocated NATO's adoption of a "no first use" declaratory policy.

The European version of flexible response (through 1989 at least) would have the Alliance maintain "sufficient" conventional military forces to defend forward in the theater but rely on the threat and possible use of nuclear weapons, escalation control through selective use of nuclear options, in the event NATO's defenses were collapsing under the pressure of battle. The objectives of the selective use would be to convince the Soviets that escalation might not be controllable (with the clear threat of the use of central systems) and to terminate the conflict, restoring the integrity of Alliance territory *status quo ante bellum*. Any strategy, policy, technology, tactic, or weapons system that contributed to this end was viewed by NATO members as desirable; any that did not, undesirable. For most Europeans, SDI fell in the latter category; after all, the president's intention was that SDI would change the status quo.

In the view of most NATO nations, US deployment (not research) of defenses against ballistic missiles would undermine Alliance strategy by causing the Soviets to increase their own efforts in defensive systems against ballistic missiles, cruise missiles, and aircraft. Such defenses would make Soviet territory and military forces less vulnerable to strikes from the West. Flexible response would lose some of its flexibility if the West had to increase the size of its nuclear strikes to ensure penetration of Soviet defenses, including possibly Soviet territory, and if Soviet responses were larger than they otherwise might be in order to get through the Alliance defenses.[6] As some commentators have pointed out, the larger the attack the less distinguishable it would be from the dimensions of a nuclear attack that would end any hope of escalation control. The premise of this argument—that deliberate escalation is possible and would be meaningful to the Soviets in the first place—is itself questionable. Yet, it has been a principal tenet of NATO's argument about flexible response—to strike near or at Soviet homeland in initial use.

EFFECTS ON NATO'S STRATEGIC CONCEPT

Since some of the actual and perceived implications of SDI for NATO have already been dealt with in earlier chapters,[7] it suffices here to concentrate on a few selected areas from the viewpoint of the allies: namely, the likely effects of SDI on war termination goals (to include the deterrence and defense premises of the proposed defense-reliant strategy), on stability in the deterrent balance between the superpowers, and on arms control or the "grand compromise" that so many wished to see happen.

War Deterrence, Warfighting, and War Termination. To convince Western Europe that strategic defenses derived from SDI technologies were worth developing and deploying both globally and regionally, the United States would have to be able to demonstrate the value of the technologies and have the know-how to integrate those technologies into effective, survivable systems. Just as important in strategy terms, however, the United States would also have to make a convincing case demonstrating how effective defenses against ballistic missiles would contribute to deterrence of war of any kind and if deterrence were to fail, to fighting the war, to limiting damage to NATO territory, and to terminating the war on favorable terms.

In sum, defenses would have to enhance NATO's deterrent strategy and improve security in Europe. Arguments justifying missile defenses through their potential to increase the survivability of US nuclear assets outside Europe or to reduce the vulnerability of US territory and population to ballistic missile attack without a coincident defense in Europe would do little to persuade Europeans of SDI's value. In fact, such points have the opposite effect. Where the United States stresses survivable forces, NATO Europe stresses vulnerable populations in the deterrent equation.

Accidental Launch: Damage Limitation and Retaliation. At one end of the spectrum, it remains to be demonstrated convincingly before any decision to deploy that SDI-derived defenses against ballistic missiles would be valuable in terms of military effectiveness and costs in limiting damage at least to key targets, civilian or military, in the event of an enemy's accidental launch of one or a small number of ballistic missiles. The attacking missiles could, of course, be armed with conventional, chemical, or nuclear warheads. In the end, the cost for protection against accidental launch—how much insurance is affordable?—will be as important a factor as the feasibility of the technologies in determining whether to deploy the systems.

The term *protection* has been used variously and equivocally to mean defense against terrorist attack with ballistic missiles (highly unlikely), the unauthorized but deliberate use of ballistic missiles (e.g., a rogue commander of a launch control center), the act of an unstable leader in a third-world nation, and launch because of human or mechanical error. On the detonation end of the warhead, such distinctions may be less crisp but nonetheless are meaningful for leaders to assess the character of any offensive retaliation. Defense, by contrast, is agnostic as to motive or to accident. However the missile or missiles were released, they would look the same to the defensive systems that would have to "kill" them. The intention of the attacker, of course, is relevant to the offense equation in determining the proper retaliatory response.

In the aftermath of the Reykjavik summit, President Reagan suggested that ballistic missile defenses, perhaps less robust than the systems required now, would still be needed even after an agreement to eliminate ballistic missiles from the arsenals of both superpowers. Such defenses could protect against cheating on arms control agreements, third party use, and accidents. While the United States places this rationale in the context of its global responsibilities, by late 1987 the European allies were much less persuaded of the need for any ballistic missile defenses in the light of the INF Treaty, banning intermediate-range ballistic missiles. With the inclusion of the SS–12/22, the SS–23, SS–20, and older missiles in the treaty ban, European interest in ballistic missile defenses in the theater was much less urgent than it had been earlier, particularly for the Federal Republic of Germany.

Early Deployment: Terminal Defenses. For a variety of budget, technology, and strategy reasons, a number of Atlanticists such as Senator Sam Nunn have rejected any possibility for deployment of "thoroughly reliable" ballistic missile defenses of the territory of the United States and its allies. However, some among the NATO advocates, as well as the supporters of ballistic missile defenses (BMD) in general, increasingly favor work on accidental launch protection. Needless to say, SDI advocates know that many of the terminal defenses envisioned in schemes and architectures for accidental launch protection are not what SDI has had as its primary interest. For the most part, SDI's focus has been space-based global defenses to protect against attack by strategic ballistic missiles, from the boost phase through all other phases of the trajectory. From the outset of the research, SDI advocates have been vigilant lest the program be captured by and identified with the ballistic missile terminal defense efforts of the mid-1960s, efforts carried forward since then in the US Army's research programs.

By mid-1988, with fewer political and military leaders in the United States and in Europe supporting deployment of any type of ballistic missile defenses, SDI advocates were willing to accept help from whatever quarter. The search for friends was especially anxious in 1988 when Governor Dukakis was advocating the end of SDI in favor of a new Conventional Defense Initiative (CDI) sponsored by Congress. Deployment of terminal defenses could be done at one site with one hundred interceptors, without any change to the ABM Treaty. However, since additional sites and possibly hundreds more interceptors would be needed for effective terminal defenses even against accidental launch, there eventually would have to be modification or abrogation of the ABM Treaty.

Contribution to Defense. In terms of strategy, analyses have yet to demonstrate how theater defenses protecting key military assets—nuclear storage sites, POMCUS sites, airfields (particularly bases for dual-capable aircraft), troop concentrations, and C^3I locations—would relate to strategic missile defenses. The recommendation of the Hoffman report of October 1983 was to begin deployment of missile defenses with "intermediate options" in the theater as the "preferred path" to achieving the president's goal. Hoffman's recommendations on theater systems matched up well with the early commissioning of regional architecture studies. However logical Hoffman was in drawing his conclusions, the idea that NATO European nations would be first to deploy anything so controversial as ballistic missile defenses was out of the realm of the likely.

In the context of the Persian Gulf, Israeli, Japanese, and the Korean theaters in particular, terminal defenses against short- and intermediate-range ballistic missiles, however armed, have grown much larger since 1983 with the proliferation of ballistic missiles. In the European theater, however, the threat has decreased. Europeans simply are not enough concerned about the threat from short-range SS–21 missiles and follow-on ballistic missiles of less than 300 miles to warrant a program to field missile defenses—even when the threat from aircraft and tactical missiles is thrown into the equation.

Contribution to Deterrence: Escalation Control and Coupling. A negative effect of ballistic missile defenses might be a lessening of the deterrent value of some of NATO's limited nuclear options. These options, contrived for the Alliance to retain control of nuclear escalation, were designed to cause the enemy to reassess the possibility for achievement of his war goals and stop the war because of that reassessment and eventually preserve the territory of the Alliance intact.[8] Another important intention of limited options has been to keep the security of the United States closely linked with that of NATO Europe and provide the president with more alternatives than an all-out strike on the Soviet homeland and the consequences thereof.[9]

Setting aside the question of whether concepts of deliberate escalation, escalation control, and escalation dominance have now or have ever had any validity, critics of SDI have suggested that the credibility of flexible response in actual war "may affect its value as a deterrence strategy less than one may think."[10] As Georges G. Fricaud-Chagnaud said, "For deterrence to work, there is absolutely no need for the aggressor to be *sure* that the victim will decide to respond; faced with the enormity of the risk all that is necessary is for him *not to be sure* of the opposite, that is to say, that the risk of response will not be nil."[11]

Nods about disarmament to the contrary, European leaders (to include most of those who have held senior political positions since 1983) have not shared President Reagan's vision of a world free of nuclear weapons or even of a world less ultimately dependent on nuclear weapons for deterrence. Support for reduced numbers of superpower nuclear weapons has not altered the Alliance commitment to nuclear weapons as the principal (perhaps only) means to deter conventional and nuclear war in Europe. In the view of many defense commentators on both sides of the Atlantic, conventional deterrence is a myth; the existence of nuclear weapons, and little else, is what has deterred war in Europe since World War II.

The history of war in Europe argues that the presence of well-armed states vying for power leads not to a standoff but to conflict. General Gallois, the enunciator of France's *force de frappe* and a principal adviser in the evolution of French nuclear thinking, in 1963 summed up a soldier's views on conventional defense and deterrence: "Most military authorities are convinced that a conventional defence of Europe is no longer feasible and that a nuclear withdrawal [of the United States] would seal Europe's fate."[12]

In the 1960s, the views of McNamara and of a US Government preoccupied with Vietnam differed radically from those of Gallois and his supporters. As Lawrence Freedman commented, McNamara had "a far better case than the Europeans would accept; a conventional defence probably was feasible and in the event of a failure in deterrence would still be preferable for the inhabitants of European NATO." According to Freedman, the debate "really did not turn on strategic analysis but on the political realities in the 1960s. It was not about the best way to deter the Russians but about the best way to tie the United States to Europe; ... in conditions of relative peace and stability [the Europeans] saw no need to tamper with any aspect of the *status quo*."[13]

Samuel Huntington's concepts of conventional deterrence and retaliation notwithstanding,[14] the view from European leaders (not their publics) affirms that nuclear weapons remain the most important force for deterrence of war of any kind. This deterrent continues to be a much cheaper alternative than a very large military establishment. From the foreign ministers' North Atlantic Council meeting in Lisbon in 1952 to the present, the Alliance has never been willing to come near maintaining the conventional power the military authorities deemed necessary (rightly or wrongly) to balance the strength of the Warsaw Pact.[15] Fortunately, the unraveling of the Warsaw Pact renders the point moot. It was not a happy prospect for Europeans to understand that strategic defenses would attempt to negate the value of nuclear forces that have served to deter conventional war as well as nuclear war.

DETERRENCE VERSUS DEFENSE

The SDI debate centers on the distinction between deterrence and defense. On the one hand, defense experts such as former Undersecretary of Defense for Policy Fred Ikle, Albert Wohlstetter, and Fred Hoffman believe that effective defensive systems can play, in Hoffman's words, "an essential role in reducing reliance on threats of massive destruction that are increasingly hollow and morally unacceptable."[16] On the other hand, there are those, particularly in Europe, who do not want to eliminate or even to reduce reliance on nuclear weapons and who do not believe the threat is hollow and immoral. Within the latter group, some believe the threat is moral precisely because it is hollow; if the threat were not a hollow bluff, it would be immoral to threaten to do that which it would be immoral to do. The allies for the most part believe that it would be irresponsible and romantic to think that strategic defenses could return the West to a pre-nuclear world.

From the spring of 1985, discussions about SDI among NATO ambassadors in Brussels went nowhere—to the relief of most nations. Given the early stages of the research, the real danger to Alliance policy lay in the ambassadors' sharing at their luncheons not only the Béarnaise sauce but also personal flights of imagination about the potentially wondrous contributions or the potentially disastrous consequences of strategic defenses. Rigorous assessments and realistic expectations were just not on the menu because no nation, including the United States, had done its homework and because there were not enough research results to display certainty.

In the United States, supporters of strategic defenses held that "a satisfactory deterrent" needs "a combination of more discriminating and effective offensive systems to respond to enemy attacks," as well as "defensive systems to deny the achievement of enemy attack objectives."[17] The view that both offensive and defensive strategic forces—for retaliation ("to avenge") and for denial ("to save lives")—are needed was a different judgment from President Reagan's initial challenge of finding a "better way."

Even given the carefully nuanced positions of those subscribing to "discriminate deterrence," the allies insisted that the burden of deterrence rest with offensive nuclear forces, forces that could reach the Soviet homeland and guarantee the coupling of the United States to Europe. The more the security of the United States itself could be tied into the direct defense of Europe the better from their viewpoint. The likelihood that escalation would get out of control, not that it would stay under control, was the kind of deterrence allies preferred. What would not be attractive was a world where the value of ballistic missiles was less than at present. Also undesirable would be a world in which the Alliance intended to prevent the enemy from attaining his war aims by having the capability to deny him those objectives rather than solely by threatening to repel and to punish him, even with nuclear weapons, in the event of his attack.

Where the debate will lead is as yet unclear. However, the issue of the efficacy of NATO's strategy for the 1990s and beyond will not go away even if SDI turns out to be merely an unfinished chapter in the search for a better way and in the evolution toward a new strategy for the Alliance. Just asking the questions about deterrence that President Reagan did in the spring of 1983 was enough for the allies to look furtively to their own futures.

Whether one looks at the British Labour party's defense policies in the 1987 national elections, the Franco-German defense discussions leading to closer cooperation between their military forces, the security issues in major US political party platforms in 1988 election campaigns, or the thrust of the "defensive defense" movements in Denmark and elsewhere in Europe, common themes emerge. The Alliance should maintain strong but reduced conventional forces, should decrease its dependence on early use of nuclear weapons, should press for deep reductions in the nuclear arsenals of both superpowers, and should remain a defensive Alliance. Alliance members would cast a disapproving look on any new "offensive" capability, especially any capability that might slow the movement toward a "new political arrangement in Europe." Most European leaders include SDI in this category.

With the Alliance questioning the relationship between offensive and defensive forces and with the United States debating its partners about how much more of the defense "burden" the "European pillar" should assume, additional pressures built to look again at the adequacy of NATO's strategic concept. As Samuel Huntington pointed out several years ago, "In its current formulation, flexible response is seen as inadequate by the strategists, unsupportable by the public, and, one must assume, increasingly incredible by the Soviets."[18]

RESPONSES TO STRATEGIC DEFENSES

Lawrence Freedman identified several categories of responses to the nuclear age that political leaders, strategists, and others have made during the past thirty years. With some adaptation, these categories of responses also provide useful descriptions of attitudes towards strategic defenses. In the first category are those who have attempted to keep nuclear war as destructive as imaginable in order to make "total war appear a greater folly than ever before."[19] In their pastoral letter of 1983, the Roman Catholic bishops in the United States strove to reinforce the tenet that there should be a strong presumption and barrier against the use of nuclear weapons. Robert McNamara and others who continue to support "assured destruction" bear allegiance to this category. Defenses that would reduce homeland vulnerability, that would protect too large a portion of the nuclear arsenal or of the nation, that would thereby bolster incentives for striking first, or that would restore superiority in conjunction with offensive forces—all would have no role in this strategic approach.

Another response pattern has been to "search for a way to deny an enemy the benefits of" the destructive potential of his nuclear arsenal "by devising either an effective defence or a form of first strike that could eliminate the enemy's capacity to retaliate." Freedman believes that this approach has been "futile" but may return in the future "perhaps inspired by new technological developments."[20] This approach, some fear, would be a search for superiority in strategic forces, a superiority the United States has not enjoyed since the early 1970s.

Prime Minister Thatcher was concerned enough in December 1984 that the United States was attempting to regain that superiority through SDI that she got President Reagan to agree that this was not SDI's purpose.[21] At least on three occasions, one right after the original speech in March 1983, Reagan publicly recognized that the Soviets might indeed believe the United States was seeking superiority through SDI, and he offered to share the technology—in exchange for cooperation on arms reductions and deployment schedules—but not free of charge. These offers, which the president almost alone thought could be worked out by cooperation, were met with guffaws by the defense establishments of both alliances, with quieter laughter up the sleeves of diplomats involved in arms control, and with nervous tittering by SDI zealots who thought this must be a wonderful trick. For the then National Security Advisor McFarlane, this encore would have complemented what he is reported to have called "the greatest sting operation in history." Those not finding the offer ludicrous include some senior Soviet officials and military officers who recognize the significance of proliferation of ballistic missile technology across the globe.

Another response has been to deny or to lessen the catastrophic destructiveness of nuclear weapons and "to contrive to develop types of weapons and tactics for their use which minimize their destructive power."[22] This response category comprehends attempts to create low-yield, highly accurate nuclear weapons for use against military targets with little or no collateral damage. As Freedman assesses this approach, however, "it is doubtful that there has been any significant success in breaking the popular association between any nuclear use and utter catastrophe."[23] In this category fit many of the principal contributors to Fred Ikle's and Albert Wohlstetter's *Discriminate Deterrence: Report of the Commission on Integrated Long-Term Strategy.*

As Iklé and Wohlstetter report, "To deter more plausible Soviet nuclear attacks, however, we also need survivable forces that could respond with discriminating attacks against military targets."[24] The report unabashedly endorses not only SDI research but also early deployment of missile defenses, just as Hoffman had recommended in the fall of 1983. "We should recognize that a limited initial deployment of ballistic missile defenses can be of value for several important contingencies." (Even partial, thin defenses can reduce an attacker's confidence, may be effective against missile attacks by minor powers, and may prevent larger catastrophes in the event of an accidental missile launch.) The commission favored a mix of strategic defenses (particularly against cruise missiles and ballistic missiles) and offensive nuclear forces as survivable as reasonably possible. This approach to deployment, which was the one (with the exception of early deployment) followed by the Services, the JCS, and the Office of the Secretary of Defense, was not the same as the president's. In the original vision, President Reagan looked forward in the not too distant future—ten years plus a bit in the aborted Iceland summit formula—to a time without nuclear weapons.

In an interpretation that Freedman would certainly find a distortion, the vision of President Reagan, often derided as romantic folly, matches well in intention with Freedman's own concluding remarks in the final chapter of *The Evolution of Nuclear Strategy*:

> Nonetheless we ought to be disturbed by the permanence of nuclear arsenals, having an entrenched position in the international order, ... to believe that this

can go on indefinitely without major disaster requires an optimism unjustified by any historical or political perspective.[25]

Freedman, too, laments the sad commentary on the human condition that stability depends "on something that is more the antithesis of strategy than its apotheosis": namely, threats that things may get out of hand. "Is there a better way?" is still the question to be answered.

STRATEGIC STABILITY

Another open issue concerns SDI's potential to enhance trust and to reinforce strategic stability between the United States and the Soviet Union.[26] In order to be valuable to NATO allies, defenses against ballistic missiles must contribute to stability (a) by decreasing incentives for a first strike, (b) by strengthening confidence between the superpowers in their capabilities for retaliatory strikes, and (c) by acknowledging, or at least not undermining, the Alliance's political strategy.

Supporters of SDI argue that defenses against ballistic missiles would help bolster disincentives for striking first with central systems. This consequence from defenses should result whether in peacetime (the imaginative "bolt-out-of-the-blue" scenarios) or in transition during a crisis from peace to war (managing the crisis under "use them or lose them" conditions for nuclear weapons). Moreover, if deterrence fails, missile defenses would help in a battle waged with conventional forces—the NATO scenario in which the Alliance military authorities may ask for use of nuclear weapons for selective options after a week or so if the battle is going poorly.

Lacking among other things adequate ammunition, spare parts, and early arriving reinforcing forces, recent SACEURs have anticipated that events would unfold in this way. What the new assumptions will be, given events in Eastern Europe, remains to be seen.

While bolstering strategic stability, deployed defenses must not contribute, it is argued, to instability in the "arms race" (a poor but persistent analogy describing both the competitive and evolutionary acquisition of military systems by the superpowers and by their respective alliances). Among other issues, one of the least examined consequences of deployed defenses would concern the importance of building and reinforcing self-protection capabilities (e.g., Patriot II's modifications) and the premium on offensive forces to suppress the enemy's defenses against ballistic missiles, as well as his defenses against aircraft and cruise missiles. Once effective defenses were in place, for example, ICBMs and SLBMs would face problems they do not have today except around Moscow—penetrating to the targets—problems similar to those that long-range bombers and cruise missiles already face against thick air defenses in the Soviet Union which protect avenues of approach to selected targets.

In this context, the purposes of defense suppression may be, *inter alia*, to enhance the chances for success of a strike by clearing the way and to remove an enemy's expectation of success for his first-strike capability by denying him defenses against a ragged retaliatory strike; the purpose of defense suppression may also be to help ensure the effectiveness of a retaliatory strike and to provide one countermeasure among many to the enemy's deployment of defenses against ballistic missiles.

Defense suppression forces, no doubt, will have an increasing role in superpower arsenals as the numbers of nuclear weapons decrease through arms control agreements and as the remaining offensive forces rely for their deterrent effect, even more than at present, on assurance of their survival through defensive measures. In the Rand analysis entitled *Strategic Defenses and the Transition to Assured Survival*, Glenn Kent and Randall DeValk accord strategic defense suppression forces the same level of importance as strategic offensive and defensive forces; in their words, "In a future era of effective nationwide defenses, defense-suppression forces will assume much more prominence and, indeed, may become the dominant force."[27]

STABILITY: A PRISM OF CONCEPTS

What the term *stability* means depends on the assumptions one brings to the concepts of deterrence and defense. The difficulty arises not with the term's connotations which, as with a word like *peace*, are nearly always positive. Rather, equivocation occurs in the multiple meanings of the word in debates over whether this or that arms control position, this or that weapon, or even this or that strategy is *stabilizing* or *destabilizing*.

Unfortunately, the qualities of *stability* and *instability*—metaphors to describe a judgment about the balance of deterrent forces, political and military—are often used as though backed by objective, scientific, or mathematical "facts" whose calculation would yield certain truths. However, as Freedman has cautioned, in the nuclear age with its remarkable constancy in the relations between political powers, "the political framework has been taken too much for granted and strategic studies have become infatuated with the microscopic analysis of military technology and the acquisition of equipment by the forces of both sides." The literature on strategy "abounds with calculations and graphs and matrices ... as if this is the real stuff of strategy."[28]

The "real stuff of strategy" has less to do with mathematical formulae about nuclear arsenals than it does with political assessments of the world situation and perceptions of truths, half-truths, and untruths as much as with "facts" themselves. The "real stuff" includes the psychology of deterrence (encompassing the very existence of nuclear arsenals capable of epic destruction); the analysis of national objectives and interests; and, even more basic, the human emotions and values of world leaders, their nations, and their peoples as they choose survival over destruction, bluff over brutality, and peace over war.

If the primal calculus in "choosing life over death" in the face of potential nuclear war were ever to break down in the political relations between the superpowers, there would indeed be the first true "instability" in the deterrent relationship in the atomic age. The elaborations of "stability" by strategic high priests are nothing other than conceptual constructs that have about them the smudges of the dot matrix printer.

With the threat of potential annihilation hanging in the strategic balance, the superpowers have been judicious, circumspect, and tolerant in gingerly but deliberately managing the nuclear balance and in equilibrating its components. In the history and psychology of nuclear deterrence, nothing suggests that the superpowers would be any less judicious in integrating strategic defenses against ballistic missiles over decades into the deterrent equation than before; they were chary, for example, with ballistic missiles themselves in the late 1950s, with multiple independently targetable reentry vehicles (MIRVs) in the early 1970s, and more recently with cruise missiles of all varieties (missiles which may well cause more difficulties for the military balance than strategic defenses ever could).

Common sense and plain talk have more to do, finally, with judgments about "stability" than the calculation, for example, of the probability of kill for a specific weapons system. Calculation has importance for any number of third order reasons, but it can never be more than a single factor brought to judgments about deterrence. Although calculation may require ingenious algebraic and statistical insights to quantify and compare various aspects of the nuclear arsenals, such mental gymnastics are always preliminary, are metaphors themselves, and, no matter how clever, are always incapable of substituting for thinking through first principles and for assessing political events that hold the center of deterrence together.

Depending on one's understanding of the tenets of deterrence, judgments about stability and instability will vary considerably. For example, in the decade or so in which the United States had unambiguous superiority over the Soviet Union in nuclear weapons and, just as important, in the means to deliver those weapons, the "atomic" relationship no doubt appeared stable from American and West European eyes. At the same time, the "conventional" relationship was clearly volatile (witness, for example, events in East Europe, Greece, Turkey, and Korea). In fact, with clear superiority, not much thought was given to the "atomic" part of the relationship until about 1955.[29]

From the viewpoint of those yearning for US superiority (European nations, at least the current governments, would not be in this category), the strategic situation has been "unstable" since the Soviet development of the hydrogen bomb and the arrival of ballistic missiles to deliver nuclear weapons. What would be most worrisome to the Europeans about superiority, even in the hands of the United States, would be the transitional phases that the alliances and the superpowers would have to manage and endure along the way toward a new deterrent relationship. Of course, neither superpower would allow the other significant nuclear superiority, Dr. Kissinger's famous question notwithstanding.[30]

For those subscribing to the concept of "mutual assured destruction," as well as for those more narrowly subscribing to the NATO theater component of that concept, strategic defenses could indeed be potentially "destabilizing." After all, the advertised purposes of such defenses are

(a) to make the United States, possibly both superpowers, and potentially the allies less vulnerable to nuclear strikes (at least less vulnerable to ballistic missile attacks) by defending against incoming missiles;

(b) to limit damage in the event of attack (and even to prevent any damage in the case of small attacks launched by second power nations or by accident); and

(c) to make retaliatory forces more survivable.

With US strategic forces (particularly land-based ICBMs) defended and with the capability to deny the Soviets any certainty in achieving war objectives in a ballistic missile first stike, the proponents of strategic defenses argue that the deterrent relationship would be more "stable"—or so their brief reads.

By contrast, apologists of MAD and protectors of the apotheosis of MAD in the ABM Treaty of 1972 argue that only destabilizing consequences will result from deployment of comprehensive strategic defenses. A world with ballistic missile defenses, in short, will be a far more dangerous place in which to live with nuclear weapons. This case focuses on four arguments:

(a) the probability of nuclear war would increase because of swelling incentives for striking first;

(b) given the comprehensive strategic offensive programs under way, there is no problem today with excessive vulnerability or lack of destructive capability in the second-strike forces in the US nuclear arsenal (there is, therefore, nothing to "fix");

(c) the deployment of strategic defenses would introduce vulnerabilities and cause yet another round of measure and countermeasure in the acquisition of defense suppression forces and strategic defenses; and

(d) strategic defense research provides leverage in getting the Soviets to reduce their overwhelming advantages in land-based ballistic missiles.

First- and Second-Strike Stability. The efficacy (that is, the survivability, capability, and credibility) of second-strike forces in the nuclear arsenals of the superpowers is the core of strategic deterrence, at least from the viewpoint of most American and Alliance strategists of the past two decades. Said another way, advocates of MAD try to strengthen second-strike capabilities and to preclude any increase in first-strike advantages on the part of either superpower. However, it is doubtful whether the Soviet Union would ever regard small, survivable, retaliatory forces as sufficient for its strategic relationship with the United States; to what extent that matters is itself open to question.

Through strategic defenses, arms control reductions, and offensive modernization the United States would avowedly be attempting to strengthen the survivability of its second-strike forces with fewer warheads by making them less vulnerable (e.g., through ballistic missile defenses, mobility, deception, improved C^3, and hardening), as well as more discriminate and more capable (e.g., increased capability for "prompt, hard-target kill" and defense suppression). However, this invigorated but smaller nuclear force in combination with strategic defenses may look to Soviets and allies like a first-strike force aimed at the center of gravity of Soviet military power and Soviet status as a superpower—namely, Soviet land-based ICBMs. It is the combination of strategic defenses, improved offensive forces and increased defense suppression efforts—and not any one element—that concerns proponents of assured destruction. In other words, MAD advocates might agree that it is prudent to improve the survivability and the capability of certain US offensive nuclear forces but not to the extent that would make Soviet forces correspondingly so vulnerable or so incapable that incentives to "go first" would become great.

In the rubrics of strategic liturgists, a state of first-strike stability exists when neither side has the incentive to launch a disarming first strike on the other's strategic forces. That is, neither side sees considerable advantage in striking first nor is "pressed to launch a first strike in order to avoid the far worse consequences of going second."[31] Some argue that even if ballistic missile strategic defenses were deployed in a symmetrical way by both superpowers to protect nuclear forces, incentives for the Soviets to "go first" would be greater than in today's situation of mutual vulnerability.[32] The proposition, as Kent and DeValk assert, would be even more certain were the Soviet Union to deploy extensive strategic defenses alone.

As is known from public sources, the Soviets have an overwhelming advantage in the number of highly capable ICBM warheads. (About 5,000 of theirs have the capability for successful attack against the 1,000 US silos; this compares to roughly 1,500 US ICBM warheads, out of some 2,000 plus, capable of attacking about 1,400 hardened Soviet ICBM silos.)[33] Assuming the Soviets strike each US silo with two warheads and with strategic defenses to protect the remaining three thousand or so Soviet ICBM warheads against US retaliatory attack (principally with SLBM warheads), the Soviets might calculate a significant advantage and decide the United States would not risk response.

Keith Payne, an early advocate of strategic defenses in his *Strategic Defense: "Star Wars" in Perspective*, replied to the collection of points that critics of SDI have emphasized in their "stability" and "instability" arguments.[34] In discussion of instability, Payne focused on the persistent criticism that ballistic missile defenses would be "leaky" (especially during any transition). Such defenses, it was argued, could cause instability. From a Soviet viewpoint, for example, the United States would be seen to be better off if it struck first in a crisis and allowed its otherwise "leaky" defenses to be much more reliable in defending against a Soviet retaliatory strike. As attenuated as such hypothesizing can get from political reality, the key question remains: even given technological successes, can defenses ever contribute to stability?

Many strategists think defenses cannot contribute to stability, believing that successful SDI research does not bear on the issue of stability and that even "leaky" defenses would be prohibitively expensive and finally useless. Because that appears to be the working assumption of so many critics—including the Union of Concerned Scientists and the "assured destruction" proponents, advocates of defenses cannot persuade opponents with the successes and promise of the research itself. In political terms, the argument for strategic defenses cannot be won against those irrevocably committed to these premises. For, as Payne points out, some have asserted that defense of US retaliatory forces, even if effective, would not bolster deterrence: "What would be gained in increased force survivability would be lost in decreased force effectiveness [in that the Soviets would also deploy such defenses]."[35] The Soviets, of course, could deploy defenses regardless of any US decisions.

No demonstration of Soviet "extended air defense" efforts (anti-aircraft, anti-missile, and anti-satellite) has yet convinced SDI's critics that the Soviets look upon defenses as a continuum with offensive forces and that no treaty, not even the ABM Treaty, has changed Soviet views on the role of defenses in winning war.[36] As the distinctions between defensive forces become blurred, with SA–10s and SA–12s already capable against some types of ballistic and cruise missiles, the question of what constitutes territorial defenses will be much less clear and will render some of today's issues about the ABM Treaty moot.[37]

Defended US ICBM forces (even apart from the potential defense of other assets such as command and control centers) would give the Soviets considerable pause. The US ICBMs could be used preferentially against the remaining, defended Soviet ICBMs or against other key targets such as massed troops, leadership, and other assets. For that matter, the Soviets anticipate attacks against their cities, witness the elaborate civil defenses, along with the continuity of government measures to protect Communist leadership, deep underground shelters, and so forth. Moreover, there is every historical and military reason to doubt that the Soviet Union would attack the US defensive forces, especially space-based elements like sensors and battle-management capabilities of the type the Soviets themselves are researching. The Soviets never attacked the US deployment of intelligence assets or even the offensive forces that threaten massive destruction of the Soviet Union; they knew the consequences of such an act. *A fortiori*, there is little in Soviet military theory and nothing in the Russian character to suggest that they would attack space-based components of strategic defenses selectively with any hope that the act would not lead to war just as surely as would an attack at sea or in the air or on land.

The point remains that defenses, even of limited effectiveness in protecting silos and strategic assets other than populations, will have major effects on military capabilities, on credibility, and on the deterrent value of NATO forces. Whether those effects are desirable and obtainable for less cost than through other means are questions political leaders, military authorities, arms controllers, and the public would have to decide. SDI's task was to help answer those questions.

The pace and direction of Soviet efforts in defenses may be affected by the pace and direction of the US research. However, Soviet decisions on deployment will ultimately depend on the prime Soviet national interest—the protection of the homeland. If the Soviets believed that defenses would protect the party and government leadership and would limit damage to the homeland, there is little doubt what the decision would be. They made that decision twice before—within the limitations of the ABM Treaty—to deploy missile interceptors for the protection of Moscow. If the Soviets decided to deploy a nationwide system, they would be a lot further along than the Alliance nations could be for years.

Arms Race Stability. Another of the arguments against "prohibitively expensive" defenses is that they will cause or "fuel" (itself a political metaphor) a costly and futile arms race, in the same way that development of ballistic missiles and later of MIRVs supposedly did. Whether or not there is validity to the charges,[38] the assertion of a potential arms race in "strategic" weapons, particularly space-based systems, has to be dealt with seriously. For the Bush administration to persuade the Congress to fund deployment would be an impossible task, especially since the US defense budget has been decreasing every year for the last six years and will continue to contract radically as long as Mr. Gorbachev's initiatives to reduce military pressures on Western Europe succeed. Increases could come, of course, from operations in the Persian Gulf.

The United States could not look to its NATO allies to make up the difference in funding. With pressures on Alliance members (from Congress, the major US political parties, and the public) to shoulder a fairer share of the defense burden, much political capital was used in the late 1980s to cajole the allies into doing more to strengthen their own conventional defense forces—to include US appointment of an ambassador for burden sharing in 1989. Such improvements, it was argued, would allow the United States to draw down the number of its stationed forces and to pursue research on long-range, defensive weapons of the type that will be needed to support a new NATO strategic concept.

Besides the defense budget itself are other unknowns: the costs of deployed missile defenses, the nature of systems developed and the schedule of their deployment over decades, Soviet efforts in offensive and defensive modernization, arms control reductions by the superpowers, the unification of Germany, the pace of dissolution of the Warsaw Pact, and other politico-military factors.

Equipment purchases are being cut and stretched and trade deficits and the national debt are problems that need rapid resolution and the Berlin Wall has come down; this is a time when it is difficult to imagine where the money could come from (estimates range wildly) to fund space-based "Brilliant

Pebbles" or even preferential or very limited terminal defenses of ICBM silos, whether or not a first phase of global strategic defenses.

When critics of SDI in Europe and in the United States raise the issue of the arms race, the argument does not hinge solely on the costs of the research (funded by the United States alone) and on the costs of deployed systems (the major part of which would be borne by the United States). The argument, rather, has its basis in action-reaction theses that many defense critics have assumed dominate defense planning and policy; that is, the Soviets react to the US development of military systems by moving ahead with their own, either the acceleration thereof or *de novo* in response. Also, the most contentious of these assertions is an implication that the Soviets would not be funding these weapons, were it not for the United States. While demonstrably false, nonetheless the idea plays well with some audiences in Europe and in the United States.

Colin S. Gray, one of the few strategists who supported SDI from its outset, described the situation correctly in a 1985 article in *Survival*. While SDI and a defensive transition can change the nature of deterrence, "in and of itself, [SDI] cannot arrest the arms competition. The 'last move' in that competition must be political, not military-technological."[39] However, as Gray pointed out, "[other than strategic defenses,] there are no attractive alternative paths to greater security."[40] Evolution and competition in the offensive systems, along with hard-fought arms control reductions, offer little more than variations on the same balance of terror in perpetuity. History provides no comfort and proffers no certainty, as Freedman has noted, that the peculiar set of political relationships between the superpowers and between the alliances that obtain today will endlessly tolerate that competition and guarantee deterrence.

The rejection of strategic defenses does not mean there will be any less competition in offensive nuclear systems. Arms control reductions notwithstanding, the competition in those offensive systems would continue as both alliances moved toward deceptive and super-hardened basing, mobile systems, additional aim points, larger numbers of warheads, long-enduring command and control, and maneuverable warheads, as well as greater accuracies and lesser yields—developments that might make such weapons, especially if fewer in number, more "discriminate" and therefore more "credible."

In arms control, however, even limited strategic defenses could help bolster the confidence of the superpowers in areas where verification is neither possible nor reliable. If properly leveraged, plans for such defenses might help achieve deep reductions in the offensive arsenals. On the other hand, with the Soviets convinced for several years that there will be no SDI major deployment, there is no bargaining chip in SDI anyway.

For those who see no escape from the dominance of offensive nuclear weapons—the "avengers" and the "bluffers"—strategic defenses remain a distraction that would be grossly expensive, offer false hopes, and undercut security. For some theorists, even discussing the need for missile defenses in order to limit damage and to deny the enemy's objectives by destroying his attacking forces is already the failure of the deterrent threat of retaliation.

At best, for some critics, strategic defenses would be just one more way to increase survivability of retaliatory forces, and maybe not the best or the most affordable way compared to deceptive basing, mobile basing, super-hardening of silos and mobile launchers, and other ways of increasing the potential for successfully attacking with a smaller force. However, the aim of SDI was the opposite: that is, to devalue ballistic missiles so that the superpowers would reduce their number and finally eliminate them as a class of nuclear weapon.

Those who posit nuclear disarmament as the principal goal—whether for political effect or not, President Carter, President Reagan and General Secretary Gorbachev declared disarmament as their policy—are divided about whether strategic defenses would help or hinder achievement of the objective. Some (the "snuffers") want both superpowers to stop testing and producing nuclear weapons and to agree to destroy the existing weapons; others have the same goal but want the West to do so unilaterally if necessary; and still others, such as SDI advocates (the "defenders"), want strategic defenses to devalue ballistic missiles to the point where they can be safely eliminated—rendering "nuclear weapons impotent and obsolete." As Colin Gray said, only multi-layered strategic defences would allow the superpowers to endorse a very radical scale of nuclear disarmament. Decades hence, as the argument goes, the money saved for purchase of offensive systems will offset the costs for defenses.

To European leaders, however, nuclear disarmament is so far from being achievable that it is dangerous piffle to suggest replacement of NATO's strategic concept on the expectation of achieving effective missile defenses. Such defenses after all, even if perfect, would still leave Europe exposed to attack by a wide range of other nuclear weapons. The allies have never accepted the notion that a protected America would be more likely to live up to its commitments. Even were that true in the far term, many Europeans would still resist deployment of defenses because of the dangerous periods the East-West relationship would inevitably have to endure during transitions to a defense-dominant deterrence; the key to persuading the Europeans (and the Soviets) otherwise is to sketch the phases of cooperative transition and to demonstrate the advantages at each phase.

5. POTENTIAL FOR A STABLE TRANSITION

Rather than relying solely on retaliation, the "new strategic thinking" would center on effective defenses against strategic ballistic missiles. By the December 1984 visit to Washington by Prime Minister Margaret Thatcher, senior officials of the administration—whether they supported President Reagan's "vision" or not—knew that much more thinking had to be done, and done quickly, about a stable transition to the new strategy. It was one thing for the administration to prevail over home-grown (mostly out-of-office) "strategists" and "scientists" in the United States, but it was quite another to stand up to the icy questions of Prime Minister Thatcher, President Mitterrand, and even Chancellor Kohl.

CENTRAL STRATEGIC CONCEPT

In the late fall of 1984, the Soviet Union had decided to begin discussions for resumption of arms control negotiations. The Soviets had, of course, walked out of the earlier talks in the fall of 1983 in protest over the initial deployments of Pershing IIs and ground-launched cruise missiles (GLCMs). Skepticism in Europe about the concept and efficacy of strategic defenses, concern about the potential bearing of such defenses on NATO's strategy, and worry about US insistence on SDI (the "president's personal initiative")—these disturbing elements all meant from the European viewpoint that the new arms talks were doomed before they began and would fail for peripheral, unnecessary reasons.

With the exception of Prime Minister Thatcher and a few others, European leaders without hesitation would have offered the American SDI program as a concession to the Soviets, right after the opening remarks of the first plenary session of the renewed arms reductions talks. At the same time, however, the administration made a convincing case that the Soviets had returned to the bargaining table because of Allied steadfastness over the first deployments of intermediate-range nuclear missiles and because of the Soviet desire to stop SDI. American officials asserted that the Soviets deeply feared SDI and the changes it portended even while the Soviet Union was engaged with extensive research into many of the same technologies. The Soviets feared that "it" might "work"—that is, result in technological breakthroughs even if the defenses were never even deployed. Most NATO members were so pleased with the opening of the new talks that they were willing to allow the United States to credit anything, including SDI; moreover, the Europeans agreed there was a plausible case.

With these national and international factors as background, Ambassador Nitze and others in December 1984 and in January 1985, in anticipation of the Shultz-Gromyko talks, worked on a short statement that came to be known as the "Central Strategic Concept." The statement set out phases for a "period of transition to a more stable world" through non-nuclear strategic defenses. Extensive review of the concept followed in the United States and in Europe by State and Defense Department officials charged with explicating SDI to the allies. In January 1985 the then Under Secretary of State Kenneth Dam, as well as Ambassador Nitze a few weeks later, publicly outlined the "strategic concept for the next ten years." In speeches, Dam and Nitze used the agreed language to link strategic defenses to arms control, to space and defense issues in the Geneva talks, and ultimately but gingerly to the goal of a nuclear free world.

In his remarks to the Philadelphia World Affairs Council on February 20, 1985, Nitze talked of the president's determination "to do more, to look even now toward a world in which nuclear weapons have in fact been eliminated."[1] In that presentation, Nitze used wording that Dam also had used and that later showed up in speeches by other senior US officials, with variations on the theme in Reagan's own speeches:

> During the next ten years, the U.S. objective is a radical reduction in the power of existing and planned offensive nuclear arms, as well as the stabilization of the relationship between offensive and defensive nuclear arms, whether on earth or in space. We are even now looking forward to a period of transition to a more stable world, with greatly reduced levels of nuclear arms and an enhanced ability to deter war based upon an increasing contribution of non-nuclear defenses against offensive nuclear arms. This period of transition could lead to the eventual elimination of all nuclear arms, both offensive and defensive. A world free of nuclear arms is an ultimate objective to which we, the Soviet Union, and all other nations can agree.[2]

Needless to say, not "all other nations" could agree with the concept. Some found the idea of eliminating all nuclear arms to be a perilous illusion that endangered security by holding out false promises, promises more political than scientific in content.

The clarification from Ambassador Nitze that the United States envisioned "a cooperative effort with the Soviet Union" toward defenses that "might make possible eventual elimination of nuclear weapons" did not make the idea any less a pipedream for the allies. And they preferred their own pipedreams. Europeans found absurd President Reagan's suggestion that the United States and the Soviet Union might cooperate closely and share know-how if the research were to pan out a motherlode of breakthroughs. It was beyond imagining that the United States would share the cutting edge of SDI technologies with the Soviet Union. After all, as Europeans never seemed to tire of reminding US spokesmen, a scant few years earlier the Reagan administration had tried to prevent the European sale to Moscow of turbine generators; the Soviets wanted the generators, based on well-known technologies, for use in a gas pipeline from the Soviet Union to West Germany. Some in the United States had not even wanted to sell grain to the Soviets in years of bad Russian harvests. The Common Market countries believed that cooperation of this sort in the superpower context was an absurdity.

Nitze identified three phases in his "central strategic concept." Over at least the next ten years (the near term), deterrence would continue to be based on the threat of nuclear retaliation. Technology, Nitze said, provided no alternative. During that time, the United States would "press for radical reductions in the number and power" of nuclear arms (particularly SS–20s and MIRVed Soviet ICBMs), for reversal of the "erosion" in the ABM Treaty (e.g., Soviet construction of the radar at Krasnoyarsk), and for agreement with the Soviets on a transition to a new relationship "based on an increasing mix of defensive systems." Presumably agreement would come on "space and defense issues" within the arms reductions talks in Geneva.

In the next phase of the "transition period," if the SDI technologies "proved feasible," the United States at some time would begin to "place greater reliance on defensive systems for our protection and that of our allies." In the best of circumstances, this transition would occur with the cooperation of the Soviets and continued reductions by the superpowers in the offensive nuclear arsenals and with the testing, developing, and deploying of "survivable and cost-effective [non-nuclear] defenses" achieved "at a measured pace." The transition would continue perhaps *for decades*. If and when deep reductions in intercontinental and intermediate-range ballistic missiles resulted from the arms talks, other types of nuclear weapons might be added to the negotiations for possible reduction or even total elimination, including nuclear weapons of the allies.

In the third phase of Nitze's conception, the reductions of nuclear weapons would continue down to zero; in parallel, both nations would deploy "effective non-nuclear defenses." Such defenses would ensure against cheating. In this "ultimate" phase, "deterrence would be based on the ability of the defense to deny success to a potential aggressor's attack. The politico-military relationship could then be characterized as one of 'mutual assured security.'"[3] Always the realist, Nitze offered no surety that the goals he set out could ever be reached; however, he still found the "brighter vision for the future" worth the effort in the absence of other choices.

During the Reagan years, Nitze's concept was the closest the administration ever came publicly to an outline, as vague as it was, for the phases of a transition to a more defense-reliant deterrence. Haunting questions remained unanswered: what assurances were there that such a transition would bolster strategic stability along the way? Would the strategic deterrent relationship be any more secure at the end of the transition, particularly with the proliferation of nuclear systems and ballistic missile technology? Presumably competition between the superpowers would continue to ensure strategic "rough equivalence"—assuming both superpowers continued to believe that no long-term advantage were possible.

Nitze's measured statement that put completion of the transition decades into the next century meant that details of the new technologies, as well as new decisions on development and deployment, could be left for administrations in the 1990s to deal with, if indeed SDI research proved out and if strategic defenses would be cost effective. In sum, there was a lot less breathlessness in the Nitze approach than in that of SDI supporters like Dr. Edward Teller, Dr. Robert Jastrow, and Secretary Weinberger. Although Nitze later supported the so-called broad interpretation of the ABM Treaty, including the development and testing of systems based on "other physical principles" under Agreed Statement D, he was not in the forefront of those pushing for early deployment of terminal defenses.

In a March 1987 opinion piece in the *Washington Post*, for example, Nitze took issue with Kissinger's support for early deployment, saying, "I disagree strongly with Kissinger's proposal to commit now to

SDI deployment. While the research program has made great progress, it has yet to determine whether prospective defenses would be survivable. Certainly there is nothing to be gained by deploying defenses that could not survive an attack: in fact, such an action could be seriously destabilizing."[4] Nitze reasserted that defenses would also have to be "cost-effective at the margin" in order to avoid "touching off a costly and destabilizing offense-defense arms race."

CONSIDERATIONS IN ANY TRANSITION

Were the strategic balance and SDI technologies to warrant deployment of interlaced systems to defend against ballistic missiles, even then certain conditions would be needed to gain Allied support. During any extended transition, the emphasis in the development of defensive technologies and in the evolution of strategy should be on increasing the survivability of Alliance, US, and Soviet retaliatory forces, not on measures to increase the capability and numbers of missiles to destroy hardened targets such as silos. Some of the European criticism about SDI stemmed from suspicion that the United States was moving toward a combination of ballistic missiles (MX and Trident with D–5 reentry vehicles) and cruise missiles (particularly SLCMs and ACMs) that would be capable of a disarming first strike on Soviet ICBM silos and strategic command and control centers. When bolstered by strategic defenses, these first-strike forces (from an Allied viewpoint) would have regained the superiority that many Europeans believe the United States continued to seek.

It was well understood that the president of the United States would have only about thirty minutes to react and to transmit messages to ride out a strategic ballistic missile attack, to launch under warning, to launch on attack, to set up to launch retaliatory strikes after the first nuclear detonations, or to decide not to respond at all. Still, most European commentators believed that situation preferable to the "automaticity" they believed necessary for effective strategic defenses. This issue of automaticity would probably become less important over time during any transition in which the numbers of ballistic missiles were radically cut and in which defenses would evolve incrementally and slowly (if not fitfully) over decades. At no time would such defenses be on "automatic," not even in Nitze's "ultimate period" when the effects of any errant interceptions would likely be nil.

A new Alliance strategic concept is finally evolving; whatever it turns out to be, the impetus for this concept will have more to do with changing political relationships among East and West Europe, the United States, and the Soviet Union than with SDI. The new strategy will have to recognize the change in the threat, limitations of flexible response, and the need for increased credibility in the seriousness and willingness of European nations to defend themselves with less help from the United States. In other words, it would be a serious mistake for Europe to read the motives behind the attraction of strategic defenses—even apart from SDI itself—as being outside the mainstream of American political thinking in substantial parts of both main political parties and outside Soviet thinking.

Whatever character the new strategic concept has, it will have to be backed by a more credible conventional defense component made up even more of European forces, as well as by new Alliance security arrangements. The relationship between strategic defenses and increased reliance on strong conventional forces (not necessarily larger forces) for deterrence goes well beyond ballistic missile defenses. Tension over these issues would obtain even if SDI were to be cancelled or trimmed severely.

Transition Stability: Issues. Since 1967 when the Alliance adopted MC 14/3, the United States has tried mightily to deal with the fact of strategic parity (or rough equivalence or whatever other term might be used to describe the Soviet Union's catching up to the United States in nuclear weapons and in the means to deliver potentially eschatological damage). During the most dangerous days of the nuclear era, the Cuban missile crisis in 1962, the United States had "probably in the order of 7,000 warheads to maybe 400 to 500 Soviet warheads;" the overwhelming superiority those numbers represent masked the even more important imbalance in US and Soviet capability to deliver those weapons at that time.[5] Within a decade to fifteen years, that US superiority had disappeared.

Since 1962, one measure after another in the offensive nuclear arsenal was pursued and either incorporated or abandoned in an effort to keep the US and Alliance nuclear forces from losing even "rough parity" with those of the Soviets. To this end, for example, the SALT process with its achievements in the SALT I and ABM treaties of 1972 attempted to limit offensive forces and to restrict deployment of defensive forces to preclude territorial defenses.

Apparently, although the Congress did not think much of the US defensive systems, the Soviets did worry about the US potential to defend. The reality was that the United States did not find its strategic defenses effective and affordable in the early 1970s, and for that reason as much as any other the United States was willing to restrict these efforts to the laboratories. The ABM system that Congress did not like was deployed briefly in the mid-1970s to defend not against a Soviet but against a projected Chinese threat that never developed.

To re-establish symmetry in the deterrent balance, other measures included MIRVs, super-hardening of silos and C^3 centers, highly accurate and powerful missiles and warheads like MX and Trident D–5, cruise missiles (especially SLCMs), and new bombers. Still waiting in the wings for their cues are mobile ICBMs (like Midgetman) to match the mobile Soviet SS–24s and SS–25s; waiting for yet another audition, after some thirty-five or so trials, is deceptive basing (e.g., to create more aim points than the one thousand US ICBM silos, a handful of strategic submarine pens, and a score or so strategic bomber bases). To the chagrin of strategic alchemists nearly everywhere, research into defenses against ballistic missiles fell into this potential mix of options—an ingredient the knowledge of which had been thought lost, at least for a decade. Some wished devoutly that such knowledge would remain lost.

Reductions and Deployments: Symmetrical and Asymmetrical. The principal problem—obvious but often overlooked—in maintaining a long-term, stable strategic balance with the Soviet Union is the Soviet penchant for large numbers of massive ICBMs with busloads of warheads, based in various modes ranging from super-hardened silos to mobile road and rail models (even a potentially space-mobile model at one time). No wonder the Soviets squealed so loudly: SDI tries to defend against the Soviets' main advantage as a superpower.

Glenn Kent from the Rand Corporation has repeatedly pointed out since the fall of 1983 that the Soviet Union has such a large advantage in current and projected ICBMs armed with multiple warheads capable of "killing" US silos that local ballistic missile terminal defenses of the silos in the United States would not help much in increasing the survivability of US land-based missiles. If the United States is unwilling to deploy ICBMs along with "modernization of basing modes" (deceptive basing, mobile basing, or both) or if it is not possible to correct the asymmetrical advantage that the Soviets have through arms control and through modernization (e.g., more "killer" warheads at sea), the United States "must prepare to build and deploy strategic ballistic missile defense capability nearly twice as fast as the Soviet Union builds and deploys its strategic defenses."[6]

A unilateral Soviet addition of effective strategic defenses would mean the US land-based ICBM force, even if it could survive a first strike in some numbers, would not be capable of retaliating with devastating effects. In sum, the land-based ballistic missiles—with their high alert rates and reliability, great accuracy, tremendous destructive potential, and redundant command and control—would no longer be so important to the US nuclear forces.

With Soviet agreement to asymmetrical reductions in the INF Treaty and with Soviet willingness to deal with an asymmetrical balance of conventional forces favorable to them in Europe, it is less unreasonable than it would have been just a few years ago to postulate that the Soviets could find it in their interest to take asymmetrical cuts in offensive nuclear forces in exchange for agreements on a limited, cooperative, and perhaps mutual deployment of strategic defenses. With SDI no longer linked to START negotiations, it may well be possible to define and agree on the best way ahead, strategic defenses as insurance in an era of much smaller numbers of ballistic missile warheads and other nuclear weapons. As the offensive systems were drawn down, the superpowers would have decades to deploy anything like comprehensive defenses. The ABM Treaty, through a series of new protocols to be negotiated at five-year intervals, would be preserved and strengthened as the mechanism to define and limit the phases of deployments.

Like the United States, the Soviet Union never abandoned research into defensive technologies after 1972. From their viewpoint, however, every Soviet ICBM warhead that would be destroyed by US defense is a warhead the Soviets counted on and is one the United States and the Alliance, except through preemption, could not destroy without strategic defenses. That is the reality, apart from any Western theory, of Soviet calculation of the correlation of forces. As aggressors, they are more concerned with inflicting damage than with limiting damage as their primary objective, once the decision to wage war has been taken.

Glenn Kent and Randall DeValk, in their Rand study, acknowledge that "the avenue along which a stable transition [to assured survival] would be possible is fairly wide." Such a transition, however, would require not only "highly survivable" missile defenses but also measures to ensure offensive force survivability (e.g., mobility and perhaps deceptive basing to increase the number of points the Soviets would have to aim at to ensure destruction of the land-based retaliatory ICBM forces).[7] In such measured analyses can be heard echoes that make yet another pitch for MX based in a deceptive scheme. Alternatively, analysis could support mobile ICBMs of the type the Scowcroft Commission had recommended in 1983.

What remains unclear in the Rand analysis is not what has been recommended but rather what can be done in practice. What will the political consensus allow? Most recommendations proffered have already been dismissed years ago in the United States, in some cases by both political parties. There is no reason to believe that things have changed, especially in light of the Gorbachev moves domestically and internationally. The public has already rejected a large number of deceptive basing schemes and would in all likelihood reject mobility for ICBMs if that meant public deployment of the missiles on rails or on highways, probably even if deployment were strictly on military bases. Given the asymmetries in nuclear arsenals, strategic defenses by themselves (especially local terminal defenses) would not be that helpful to the deterrent equation for a long time to come, as General Glenn Kent demonstrates.

What yet needs to be done is a systematic study of the most desirable integration of strategic defense (including active and passive self-protection), strategic offense (including defense suppression), and arms control reductions. The US desires, with perhaps even Soviet acquiescence, to reduce the "destructive potential" of land-based ICBMs in an asymmetrical way through arms reductions. However, there does not appear to be, at least in public diplomacy, any consensus or scheme for how many Soviet ICBM warheads need to be eliminated and in what phases in order for territorial or at least preferential regional strategic defenses to be effective. In short, what does it take in reduced arsenals and in deployed defenses to bolster mutual assured survival?

As Paul Warnke, former head of the Arms Control and Disarmament Agency and a long-time advocate of deep arms reductions, has pointed out many times, some results from arms reductions would leave the worst of all imaginable relationships. For example, if the superpower arsenals each included 100 ICBMs with 10 "prompt, hard-target killer" warheads on each missile and if the missiles were based in 100 silos, the incentives for first strike would be extraordinarily dangerous. Each superpower would have a 10 to 1 ratio of warheads to silos if he struck first, and no capability to retaliate with ICBMs if he *was struck* first. The greater the number of potential aim points for the defender in relation to the number of weapons for the attacker, the more stable will be the ballistic missile component of nuclear deterrence and the easier would be the tasks of ballistic missile defenses.

Proposals for reductions in the range of 30 percent to 50 percent in the current arms reductions talks (START) would be helpful to the case for strategic defenses if those reductions were taken with this principle in mind. Absent a national commitment to deployed strategic defenses, US arms control positions do not rightfully reflect the best potential combinations of weapons for a transition to a defense-dominant deterrence. In other words, strategic defenses remain *a* factor in the calculations, but not *the* factor.

Competitive and Cooperative. Even in the original speech launching SDI, President Reagan was alert to the need for cooperative arrangements with the Soviet Union, for example, "to achieve major arms reductions." However, with such reductions, it would still be necessary "to rely on the specter of retaliation" for a long time to come. He also recognized that even though "defensive systems have limitations and raise certain problems and ambiguities...[the research and development efforts] could pave the way for arms control measures to eliminate the weapons themselves." With that, he launched "an effort which holds the promise of changing the course of human history." If intention rather than consequence were the measure for the efficacy of strategic defenses, the boldness of the president's claim perhaps would be better accepted by SDI's critics, to include those carrying pastoral croziers.

For better reasons than simply lassitude in "officialdom," little analysis has yet been done on potential transitional steps towards strategic defenses over the next several decades.[8] It would appear self-evident that it would be easier for the United States to proceed with Soviet cooperation than without it. The United States could concentrate on measures to force the pace and the direction of the transition, just as it has in strategic theory about deterrence, about arms control, and to some extent about weapons development since the atomic age began. However, several factors militate against cooperation, beyond mistrust that might prevent cooperation altogether and apart from the satisfaction the Soviets have felt in achieving parity in nuclear arms.

No matter how close the cooperation, the United States and the Soviet Union, after several phases of deep reductions, would each separately have to cast a wary eye on ballistic missiles of all ranges in the arsenals of other nations, minor and major. In addition to the nations today with long-range, nuclear-armed ballistic missiles, there are several other nations that already possess or will acquire before long short-range and intermediate-range ballistic missiles. Several of these nations are capable of manufacturing ballistic missiles (in some cases, to include nuclear warheads) for their own arsenals and for those of their customers.[9]

The terrorizing attacks with medium-range ballistic missiles in the Iran-Iraq war in the spring of 1988, attacks that helped devastate the Iranian fighting spirit, showed what may lie ahead. Although such missiles armed with conventional warheads, especially with improvements in high explosives and in accuracy, could do damage at long distance, the main threat includes over the next decades chemical warheads, as well as nuclear warheads. To respond to such contingencies, the superpowers would want offensive and defensive measures. Whether or not that would mean a minimal, nuclear-armed ballistic missile force for retaliation, along with strategic defenses, would need to be part of transition deliberations.

Although the Soviet Union has a unilateral advantage in both deployed ballistic missile defenses and in an operational anti-satellite (ASAT) system, the United States has elected to date not to continue with full-scale development and deployment of these capabilities. The US Army now has the lead in developing an ASAT interceptor and system, distinct from programs under the SDIO. In the past, the BMD and ASAT systems were deemed not worth the cost and not effective. With development and (more important) the deployment of defenses (either nationwide defenses or comprehensive defenses of silos), there could be no restoration of unilateral advantage. Neither superpower could allow that to happen.

Unilateral deployment of strategic defenses would constitute a turn of events that would be the most disturbing in the history of superpower competition—especially if the Soviet Union had the unilateral advantage in effective territorial defenses. With a 3 or 4 to 1 advantage the Soviets would have in "prompt, hard-target killer" warheads and with effective territorial defenses, the incentive for a first strike might be overwhelming and the deterrent balance would erode commensurately.

In a dramatic reversal of the exchanges at Glassboro between McNamara and Kosygin,[10] General Secretary Gorbachev and former Secretary of Defense Weinberger have each stated separately that he would increase the number of nuclear warheads and take other countermeasures in the event the enemy deployed strategic defenses unilaterally. Neither nation could tolerate nearly "leakproof" territorial missile defenses in the other nation for long. The French and British certainly do not want to face Soviet territorial defenses. The purpose of the ABM Treaty of 1972, of course, was to preclude just such defenses.

To avoid a competitive race to procure defensive measures, defense suppression forces, and countermeasures, the superpowers would need to cooperate in setting out phases of a transition that would leave the deterrent relationship stable at every stage. The history of superpower experience in such cooperation since World War II leaves this proposition dicey at best—even with the dazzle of the recent political changes. On the plus side, however, are the strained defense budgets; both nations (and their allies) face financial struggles, in the case of the Soviet Union in economic development and expansion and in the case of the United States in dealing with the massive federal budget deficit and unfavorable trade balances. At the time of SDI's launching, no "expert" could have imagined the changes that have occurred in the superpower relationship and in the nature of NATO and the Warsaw Pact over the past few years, especially in 1989, the year of "revolutions" in Eastern Europe.

In the best imaginable world, it is possible to calculate tremendous potential savings or cost avoidances over the next two to three decades from offensive systems forgone through mutual agreement, enough "savings" to pay for small but effective strategic defenses (to include defenses against aerodynamic (cruise missiles and aircraft) and ballistic missile threats (land-based and sea-based). The chance of such a rational scheme occurring, however, is small to nil, and not one on which to pin the wriggling hopes of Alliance security. Without cooperation in reducing the superpower arsenals radically and in agreeing on the cooperative deployment of defenses in order to ensure stability, the shift to a defensive strategy of deterrence would be too expensive and too risky. The Persian Gulf situation remains a spoiler to planning a phased evolution, given unknown budget demands.

Population Defense and Weapons Defense. From the outset of the SDI program, detractors and zealots alike have used inconsistent if not sometimes contradictory justifications to support or to criticize how strategic defenses can on the one hand "enhance" deterrence and can on the other hand help replace the current "immoral" strategy of deterrence. While paradox, ambivalence, and dramatic irony might properly characterize public and private debate and policy-making about the components of deterrent strategy, what cannot be tolerated is contradiction in the tenets of the military strategy that derives from the national strategy.

While national or grand strategy may tolerate or indeed require the oxymoronic yoking together of sometimes contrary policies and interests, military strategy cannot enjoy such tension without fragmenting and losing support (both internal and external to the troops). By its nature, military strategy needs as much clarity in policy as possible, especially since the fog of war itself will provide all the ambiguity tolerable and will be more than enough to contend with. As soon as public understanding diverges unduly from the policy that actually informs war planning and the plans themselves, strains follow inevitably and make decisions Gordian, at best, on issues concerning the credibility of the nuclear arsenal, as well as on arms control policies.

Such nuclear weapons debates get more concrete when the focus shifts away from what weapons are to be used and towards what is to be defended.

The president had asked, "Wouldn't it be better to save lives than to avenge them?" The best answer is that deterrence means it would be better not to have to do either, that is, not to have to save lives or to avenge them. However, in the event of attack, the defender would want to do both—save the lives of his own people by defending them from attack and inflict damage and punishment on the attacker to get him to desist. Needless to say, that was not a permissible answer for the public and the allies to use in response. President Reagan simply did not offer sufficient alternatives. An over-long, logically tumultuous, and finally irreconcilable debate has attempted to answer the wrong question.

At least four schools (*strains* might be a better term) of thought emerged from the nuclear weapons debate. The schools can be defined more or less by their focus on what targets would be protected (the "high heelers" of population defense and the "low heelers" of weapons defense); on when defenses could be in place (from five years to several decades, respectively the "early deployers" and the "late deployers" or even the "never deployers"); and on what defensive technologies, developed systems, and basing sites might be used (ranging from basing on US land to space-based systems in echelons).

Membership in one or the other group has implications for views on the value of the ABM Treaty, the narrow or broad interpretation of Agreed Statement D, the worth of early deployment of terminal interceptors, the requirement for comprehensive defenses, the toleration of less than "perfect" or "leaky" defenses, the need to replace NATO's strategy of flexible response, and a number of other arms control, force modernization, and strategy issues. Nearly a decade of graduate students has already feasted on the chum from the debate among interested parties on both shores of the Atlantic. Those who perhaps have benefitted most, apart from the corporations with SDI contracts, have been political cartoonists.

Cartoonist Gary Larson captures the mistrust in technology and the likelihood that a system might not "work" when needed.

THE FAR SIDE COPYRIGHT 1986 UNIVERSAL PRESS SYNDICATE.
Reprinted with permission. All rights reserved.

THE PURISTS *and* ABSOLUTISTS: ALL *and* NOTHING

In one group are those who demand or expect, depending on whether they are critic or advocate, that SDI-derived strategic defenses will be or must be perfect. That is, the defenses would have to provide "leakproof," multilayered, global defenses (space-based in most versions) against ballistic missiles in order to protect the people and soil of the United States and of its allies from attack or else they should not be deployed. Taking advantage of the exuberant claims and hopes of SDI zealots and using their claims to undermine, hector, and ridicule SDI and its supporters, critics asserted that anything less than "leakproof" defenses would not be enough and would undermine the president's vision. They took this position not because they cared for that vision but because they found it not a vision but a nightmare.

Such defenses were in any event not feasible according to a large number of "scientists" and "strategists" who lent their names to battles, pro and con, about the sagacity of a strategy of deterrence based more and more on defensive systems. If defenses were not possible, it was *a fortiori* unnecessary to have an expanded research program to investigate the technologies—or so the argument went. A more modest program—the minimalist position—would do just fine as a hedge to Soviet activities. Such a program would not tempt SDI advocates into early deployment and into abrogation of the ABM Treaty.

Also in this group, for quite different reasons, are those who have advocated an Apollo-like project to move rapidly toward deployment of extensive, space-based strategic defenses. In fact, some of those who were the original advisers to the president on SDI thought the technologies were already at hand in 1983 for the first phases of space-based defenses capable of post-boost and mid-course interceptions of warheads.

For both groups of purists, then, the idea of transition was meaningless. For those who believed the technologies were at hand and inevitable, the transition had begun long ago. For those who believed the technologies would never be available, talk of transition was an absurdity. For many who knew something of the technologies under investigation even before SDI, deployment seemed easier to think about than it did for others not cognizant of developments in ballistic missile defenses and of developments in command, control and communications. As always, the zealots probably saw more and the critics saw less than was actually there.

The deployment purists wanted to ensure that the president's "vision" would not be destroyed by bureaucrats and by apologists of mutual assured destruction (MAD). The vision had to do with returning the assurance of national survival to US hands. Unlike the current situation, survival would not depend on Soviet good will and adherence to MAD. After all, the ability to retaliate lost much of its meaning if the devastated nation would not survive even though it could retaliate. The purists wanted to get agreement to early deployment of some systems, however far from their final goal in effectiveness, in order to commit the nation to strategic defenses in principle.

Only a short time after the president launched SDI, advocates in the bureaucracies found a way to escape their disjunction, by simply asserting that defenses would not have to be perfect in order to contribute to deterrence. Defenses would increase the uncertainty in the mind of the potential attacker about the success of the attack and therefore would help deter. With this rationale in mind, the president himself could agree with Prime Minister Thatcher in December 1984 that SDI's purpose was to enhance deterrence. The designers of SDI, however—including the president—had grander purposes in mind from mid-1983 to at least mid-1986.

The arguments of both advocates and critics exhibit major flaws concerning whether strategic defenses have to be perfect in order to deter attack on and, if necessary, protect populations. Both arguments cause their own confusions through the reification of SDI defenses. That is to say, SDI is neither a single thing nor a single system—and never will be. If deployed, strategic defenses would be a series of defensive systems put in place over decades through multiple evolutionary steps. Early parts of the integrated series would no doubt be a generation or two behind later stages; various elements would be obsolescent at any one time, just as with other weapons.[11]

The series of systems would never be complete, but like other military weapons would continue to evolve based on the threat, countermeasures, defense suppression advances, and cost-effective improvements to the defenses. Moreover, at some stage, the systems would also have capabilities to defend against remotely piloted vehicles, aircraft, and cruise missiles, as well as ballistic missiles like the SS–21 that remain within the atmosphere.

Radical reductions in offensive nuclear arms would be a *sine qua non* for such defenses even to approach comprehensive effectiveness. To plot strategic defenses against current and planned forces is, although necessary for analysis, already to have missed one of the main parts of the envisioned transition to strategic defenses. Without deep reductions in offensive systems and without a measured evolution of deployed defenses, the vision could never become reality. "Well then, so much for a naive vision," say the critics.

To his credit, Lieutenant General Abrahamson stayed as close as any program director could to the original vision as the final goal for his organization. While proponents of early deployment offered their support, it was too large a price to pay if such deployment meant only land-based terminal defenses of missile silos. That was what the US Army had been researching since the early 1960s, not what SDIO was doing. Let the Army continue its work; SDIO was doing something different.

DISCRIMINATORS: OLD BUSINESS WITH A NEW MIX

Another school of thought gave a nod to the president's vision but then proceeded in a familiar approach to defense of ballistic missile silos and by extension, perhaps, defense of command and control and leadership centers. As part of the Future Security Strategy Study, for example, the Hoffman report of October 1983 began this blurring of the vision early on. The vision was of presidential making—enough said for administration officials and supporters. Had the R&D bureaucracies of the government conjured the vision, however, some in the community of policy-makers might have had a lottery for the privilege of casting the first stone. By mid-1988 the initiative for many had changed into limited defenses through the interventions and cautions of the allies, of congressional critics like Senator Nunn, of strategists ranging from Harold Brown and James Schlesinger to the "Gang of Four," and of critical reports ranging from those of the Office of Technology Assessment to the June 1988 Roman Catholic bishops' follow-up report on their earlier pastoral letter entitled "The Challenge of Peace."[12]

Even though the SDIO tried to concentrate on boost and post-boost efforts with space-based kinetic energy weapons (with directed-energy weapons to come along decades later), some approaches to missile defenses began more to resemble those of the late 1960s. That is, defenses of silos and of command and control and leadership centers. From Fred Hoffman's discussion of "intermediate options" as the "preferred path to the President's goal" and from Hoffman's suggestions for "anti-tactical missile" (ATM) options for the allies in his 1983 report and through Senator Nunn's support five years later for efforts to protect against accidental launches (ALPs) by an enemy, the focus on limited defenses has persisted.

Such limited defenses supposedly could give the United States some protection for and could defend on a preferential basis against small "accidental" launches. According to supporters of this version of strategic defenses, this goal would have a more realistic chance for achievement than the vision of population defenses in the face of massive attacks by the Soviet Union.

The "discriminators" would no doubt object vehemently to characterization of their approach toward SDI as business as usual. The principles of "discriminate deterrence," as much as these can be discerned among the assessments published to date, lead to support for a mix of highly accurate and discriminating offensive systems (with arsenals much reduced perhaps) and effective defensive systems. This mix occurs in the context of a "technological revolution in military affairs," powerful new actors on the international stage in a multi-polar world, and "a large relative decline" in the economy of the Soviet Union.[13]

This group looks to "probable revolutionary improvements" in military technology that "could fundamentally change the nature of warfare": for example, "long-range surveillance, target acquisition, and weapon delivery systems, and low-observables aircraft and missiles."[14] However, despite all the breathless discussions of the applications of technologies to military weapons systems over the next twenty-five years and despite the vision of the "strategic" capability of future conventional, non-nuclear systems, the focus remains on operational art, theaters of potential conflict, and improvements in conventional weapons. At the level of national strategy and Alliance theater strategy, nothing in the nature of deterrence itself has fundamentally changed.

EXPLOITERS AND OTHER PIGGYBACKERS

Without commitment to full-scale development and deployment of strategic ballistic missile defenses, some senior officials in the defense establishment and more widely in the administration found SDI, especially at the outset, to be a fortunate braiding of events that could be exploited for a variety of private agendas in arms control and in the modernization of forces. In this third category some participants, while choosing to lie on the same pallet, felt a need to distinguish and accommodate their various approaches.

For example, Strobe Talbott, in his May 1988 *Time* article entitled "Inside Moves," asserted that the then national security adviser to the president, Robert McFarlane, initially supported SDI as part of his "elaborate covert operation to lure the Soviets—and the President himself—into an arms control deal." According to Talbott, "in McFarlane's mind, SDI was a step toward an agreement in which the program would be limited in exchange for diminution of the Soviet offensive threat."[15]

At the east side of the arms control "apse," Talbott locates then Secretary of Defense Weinberger and Richard Perle, his assistant secretary for international security policy. Both saw SDI "as a way of spiking the wheels of the [arms control] process." To this end, they became "champions of the President's dream, in its most ambitious, least negotiable form."[16] No one who knew and dealt with them would doubt that Perle and Weinberger were clever enough to exploit all the city alleys and country paths of SDI to ensure US advantage in arms control agreements—especially if agreements could be negotiated that overwhelmingly favored the United States and did not disturb the US military build-up.

At the same time, no mortal or group of mortals could have controlled and directed the complexity in the shifting motives, the nuanced motivations, and the labyrinth of arguments for and against SDI on both sides of the Atlantic—including the maturing and ever more forthcoming Soviet reactions. That would be too much credit to give.

Perle and Weinberger seemed content enough with Ambassador Nitze's concepts for transitional phases—the concepts were vague and sufficiently in the future not to cause any problem. They became irritated over time, however, with Nitze's criterion of "cost-effectiveness at the margin" for any decision to deploy defenses; they preferred instead the concept of "affordable." That is, if the technologies become available to field effective defenses, then by definition the public would find them affordable in terms of saving lives and property. It was too late in the SDI rhetoric to break the tablets.

Among those who used SDI to carry their own agendas were defense industry leaders in Europe, Japan, Israel, and the United States. Whether lifelong zealots of strategic defenses or born-again believers of convenience in defenses against ballistic missiles, industry officials did much to make national security elites aware of the potential return from SDI research, as well as of possible future production business for those companies that were in at the outset of the program.

What industrialists understood intuitively was that whether or not anything ever came from SDI, there would be important translations of the technologies to conventional defense areas such as communications and command and control and to kinetic energy weapons. They understood that SDI would capture and exploit some of the anti-nuclear fervor in the West; defensive measures would probably be the theme of the 1990s since offensive measures had about run their course for now. Industry also knew that modernization of nuclear systems appeared stagnant (the industrial focus goes where the money is) and, finally, also knew that what was needed was to help establish a constituency in the Congress for support of strategic defenses. The easiest constituents were members whose states had SDI contracts and those who already believed in terminal defenses.

Industrial conferences sprang up, more like toadstools than mushrooms, from mid-1985 on. There was a lucrative cottage industry in sponsoring such conferences and in having the clout to get the best (read, the most influential) speakers—civilian and military government officials, scientists, academicians, strategists, and businessmen. Although strategy issues were sometimes raised, for the most part the conferences became the gathering places for the faithful and the enlightened—that is those who supported not just SDI research but also supported deployed missile defenses. After all, without deployment, the funding for SDI would eventually get harder to come by. The potential consequences of SDI for strategy were rarely if ever the main issue.

APOLLONIAN AND DIONYSIAN ARMS CONTROL

Those favoring disarmament through the arms control process might conveniently be divided first into those who do so through a rational and objective analysis of nuclear and conventional force balances vis-a-vis the threat, as well as the effects on stability at various levels of reduced arsenals. Second are those who viscerally look upon such analyses as venal, unnecessary attempts to justify reductions for nought, reductions not needing any rationale. For the latter, whose arguments are emotional more often than discursive, the justification for radical or total disarmament is self-evident and elemental.

At the top of any list of disarmers must be President Reagan's name, given the stated intentions of his SDI program for the long term and his personal negotiations on arms reductions at Reykjavik, that is, for a world without nuclear weapons. The point is not that any defense expert holds that the objective is not noble but that many believe the goal is not achievable and is therefore dangerous to offer the NATO and Soviet publics. In other words, disarmament is a chimera that will only distort and distract from the serious business of arms reductions. Nonetheless, Reagan repeatedly struck this theme of a world free of the nuclear threat.

Among the critics, some thought they saw in SDI only chicanery and public relations as President Reagan continued his military build-up; some thought they saw the clever preparation for the abrogation of the ABM Treaty, and still others thought they saw naivete of the worst sort, driven by the deadly twins of arrogance and ignorance.

BARGAINING CHIP OR CHIP ON THE SHOULDER?

While European leaders such as Prime Minister Thatcher, Chancellor Kohl, and President Mitterrand might well have understood the motives in the United States for the anti-nuclear undercurrents in the SDI program, they nonetheless had to reinforce their belief among the European body politic that nuclear weapons would not go away but would remain the central force behind, as well as the guarantee for, deterrence. To many Europeans (as well as to not a few American theorists), conventional forces have little deterrent value, absent nuclear arsenals and the threat of first use. As discussed earlier, the extravagant claims for SDI perhaps frightened the allies more than the Soviets since SDI struck at nuclear weapons in ways that supported those who were seeking the "denuclearization" of Europe through arms control, through reductions in defense spending, and through denial of modernization for nuclear forces.

What Prime Minister Thatcher agreed with President Reagan in December 1984 was in fact what happened through the beginning of the Bush administration: namely, events unfolding included an ABM Treaty intact (having withstood much pressure), continued research into defensive technologies as a hedge, no decision to deploy, and an open question about deployment in the context of negotiations on arms control reductions. The early part of the Bush administration also supported a commitment not to seek superiority through strategic defenses, SDI as enhancement of and not a replacement for deterrent strategy, and the US commitment to continue to consult closely on developments as SDI evolves. Absent a sealed agreement with the United States that nothing would happen unless Europe agreed, Europeans nonetheless by the end of 1988 were as confident as the Soviets that SDI was not to be the frightening initiative that it had been pledged to be at the outset.

The issue of transition, the Europeans thought, would never be a serious one since there would never be a decision to deploy global defenses. The United States could do what it wished about defending silos on its own territory, as long as it did not take any funds from the US commitment to Europe. In short, European timidity would control American temerity attempting through missile defenses to strike a new deterrent relationship with the Soviets and, perhaps more important, with the Europeans. One can hear the chortles of General de Gaulle, who would remind his colleagues of the indivisibility of nuclear forces and threats, and of the primacy of survival and sovereignty.

6. SDI'S BEARING ON NATO

With apologies to the grammarian H. W. Fowler, categories that Fowler developed (for an unrelated subject) circumscribe with some modification most who have thought about SDI. Observers may be divided into those who neither know nor care what SDI is in any detail, those who do not know but care very much, those who know and either condemn or approve, and those who know and distinguish.[1]

This present work is intended to be in the last category—that is, not a zealot's brief, not a critic's disprizing commentary, but an exploration of questions about the potential bearing of SDI on NATO's strategic concept. In my view, a definitive assessment of the potential consequences cannot in fact yet be done without carrying along brittle assumptions that render conclusions fragile. Simply put, more needs to be known about the technologies, about the balance of nuclear forces under a number of possible arms control regimes, about the resultant threat and order of battle, and about modernization to improve the survivability of nuclear forces. More also needs to be known about evolving political relationships between the United States and the Soviet Union (as well as between NATO and East European nations), about the proliferation of ballistic missile technology and nuclear weapons in nearly a score of nations, and about the commitment the nation might be willing to make to deploy defenses in space.

Absent such an integrative approach, a compelling case cannot be made that today would underwrite the decades-long commitment of national treasures, that would warrant the enormous contribution of the talents of scientists and strategic thinkers, and that would inform the arduous negotiations and necessary transitional phases to see global defenses through to completion. This observation is not to toll a bell for SDI, but to put deployment of space-based systems in a realistic context. The judgment is also not a charge to stand still; on the contrary, the recommendation of this study is to press forward with the research and technology efforts, as well as with arms control reductions in the ballistic missile arsenals.

MISOLOGISTS NEED NOT APPLY

Those in the first category—"who neither know nor care about SDI in any detail"—comprise the largest group by far. The superficiality that has sometimes passed for investigation since mid-1985, occasionally by otherwise "knowledgeable" commentators, precluded discursive reasoning leading to a national commitment for or against strategic defenses. The substitution of fiction, assertion, and promise for fact, analysis, and assessment helped make it impossible to reach public consensus on a way to get ahead. For the sake of harvesting of the future research results, there should be no press for commitment until there is a compelling argument that the new structure of deterrence and the new balance of forces will be stable, survivable, and secure over many decades—especially in a world of multiple centers of political power.

The majority of the US public (who are in the first category) believed that strategic defenses were already in place, in some rudimentary form at least. Most of the US population, if the polls have been correct, had no difficulty with the concept of defenses, even space-based systems, against ballistic missiles. For the public at large, the case for SDI appeared to be about very little; large segments apparently thought SDI to be no more than concentrated efforts to improve existing systems.

Military and civilian defense officials with some knowledge of missile defenses, on both sides of the Atlantic, quite rightly occupied themselves not with SDI but primarily with the preparation and modernization of the conventional and nuclear forces that underpin Alliance security. From their viewpoint, SDI was and should remain research for a long time.

For those who acquired and prepared forces to deter and to fight if necessary, it made sense to keep a wary eye on the potential of strategic defenses to consume too many defense dollars over too long a time. The burden of proof for convincing the bureaucracies remained with those who would change the strategic equation radically with deployment of missile defenses.

The public has not yet had to pass judgment on specific deployment issues.[2] However, the public would probably show considerable uneasiness over the placing of nuclear power generation systems in space, as well as a strong presumption against the deployment of weapons in orbit. Kinetic energy weapons, however, would be less contentious than nuclear weapons—especially were those space-based weapons to remain impotent until the correct enabling commands were given. In the United States, the opposition to deployed strategic defenses, not to

SDI research, was led by elites with particular political, scientific, religious, and strategic views on disarmament, the arms race, weapons in space, nuclear weapons and power generation, the importance of the ABM Treaty, and deterrence based on the threat of massive retaliation.

The category of "those who do not know but care very much"—more given to emotion than to reason—has caused most of the misunderstandings about SDI. One end of the spectrum finds the zealots for SDI, ranging from those who want strategic defenses deployed right now with technologies at hand to those who know the technologies are immature but want to proceed anyway with interim steps in order to capture the residual support of the American public for defenses against ballistic missiles. At the other end of this group are critics, ranging from those who want to stretch SDI out indefinitely because of SDI's potentially negative effects on strategy, arms reductions, and the military balance to those who would not want SDI even if the technologies were in fact available and "worked." Most in this group fervently hoped that SDI would turn out to be an aberration of the Reagan administration and that SDI would drift away in stages with the Bush presidency.

In the third category, "those who know and either approve or condemn," on the approving side from the outset, were a number of prominent scientists such as Drs. Edward Teller, Robert Jastrow, James Fletcher, George Keyworth, and Fred Seitz and General Graham, as well as office holders like then Senator Quayle and then Congressman Kemp, who knew the technologies available as well as the promise of the work in the laboratories. With some variations, however, they reached generous conclusions about how effective and how comprehensive the technologies for the missile defenses might be, as well as about how quickly such defenses could be put in place.

Among those who "know" and condemn, scientists like Hans Bethe, Richard Garwin, John Pike, and other scholars—bolstered with many statements from the Union of Concerned Scientists against the potential cost, efficacy, and strategic implications of missile defenses—did all they could to ensure that SDI remained research and that there would be no national commitment to deployment. From mid-1985 to the present, there has unfortunately sometimes been an almost "rent-a-scientist" character to what has served for a public debate about the technical and technological merits of the SDI research.

Each side of the fundamentally political arguments sought "scientific" authority for its positions. In terms of numbers, the public impression (and probably the reality) was that most of the scientific community (including "scientists" who knew little or nothing about the technologies) were opposed to deployed strategic defenses against ballistic missiles. However, they were often opposed to SDI for reasons other than strictly "scientific" ones. The important question, however, was whether such opposition mattered to the public. For the most part, the answer was no. Since the arguments of the scientists appeared to turn more on foreign and domestic policy, as well as economic, military, and strategic points, than on science, the pronouncements of such scholars, however definitive, were not enough to close the policy debate or to turn it against SDI. In fact, the public continued to support SDI even though the understanding of what "it" was remained decidedly imperfect.

Having reserved the most favorable classification for the present work, I hope this analysis has fit in with "those who know and distinguish." Neither the fish of zealots nor the fowl of critics, this study has explored the main concerns the European allies have been struggling with since President Reagan launched the initiative. These findings, not astoundingly, fall within the mainstream views in America at the end of the 1980s, namely, do not rush to deploy limited defenses:

—continue the research at or about the current level of funding ($3.0 to $3.5 billion) for a few more years;
—continue the testing of sub-components only and preserve the boundaries of the ABM Treaty;
—use the ABM Treaty as a way to negotiate measured "rules of the road," as well as incremental and transitional steps to deployment of strategic defenses (if that turns out to be the right policy derived from the research under way and from serious strategy studies);
—bring into harmony US and Allied approaches to modernization of forces, arms reductions negotiations, and strategy debates. Although a number of reviews are underway, it remains to be seen whether any current efforts at NATO or in the United States will advance the state of Alliance calibration of offense-defense relationships in a new strategic concept;

- integrate these efforts in defensive technologies into a national security strategy that will inform US policy in the multipolar world of the twenty-first century;
- carefully distinguish the "accidental" launch protection argument from the "terrorist" argument as well as from the deliberate use of ballistic missiles by hostile nations in conflict with the United States or its allies; and
- above all else, do not rush headlong to make decisions on the deployment of strategic defenses.

There is time to see if SDI is the "better way" to manage the strategic relationship, especially in the context of the revolutions of 1989 and 1990.

At the story's end, Dorothy does make it back to Kansas, and nothing can make it otherwise. But we do not yet know the final chapter on SDI. Even with all the wonders of SDI to date, however, Alliance leaders and their publics do not know whether global space-based strategic defenses will ever be effective and affordable by any reasonable standards, as well as contribute in significant ways to crisis stability and deterrence, and therefore be worthy of deployment. Despite what the zealots and the critics might wish, nothing more can be known for sure and nothing can make it otherwise—for now at least.

CONCLUDING THOUGHTS AND RECOMMENDATIONS

The policy argument for deployment of strategic defenses against ballistic missiles has not grown more persuasive to the European allies. Moreover, in terms of their own homelands, the INF Treaty that eliminated SS–20s, SS–12/22s, and SS–23s has made

Europeans feel even less threatened by ballistic missiles. Although there have been technological successes (in some instances, spectacular) in a number of SDI program elements, there has not been the rush of scientific and political support in Europe that might be expected for an idea whose time had come. In fact, at times it seemed that breakthroughs in SDI led to near breakdowns among European political leaders who preferred the status quo.

By the time President Bush came to office, Europeans were content to wait it out to see where America would take SDI. The less said, the better. At the same time, despite complaints from zealots and critics alike of SDI, there has been remarkable stability in congressional support—even if not, of course, to the level originally envisioned by the White House. Until FY 1990, there had been substantial annual increases in SDI funding even in years in which the overall DOD budget had negative real growth. (See appendix B.)

The original vision now shows the wear and tear of the abrasives applied by those who tried to make it more and less than it was. Moreover, the SDI vision suffered from insufficient funding to achieve all the goals by the early 1990s and from a lack of solid work on strategy and arms control implications. Clearly there would be no clean, early break with the Alliance's strategic concept and with the ultimate threat of assured destruction.

A reasonable approach that has evolved, however, is that over time a series of missile defense systems could be deployed incrementally and eventually laced together. The Defense Science Board Task Force Subgroup on Strategic Air Defense, in its May 1988 SDI Milestone Panel report, arrived at much the same conclusion and set out transitional steps to satisfy military requirements, with each step having its own rationale as well as the possibility of contributing to a larger system.[3] The missile defense systems, integrated as much as possible, would be of varying levels of effectiveness against threats that would themselves continue to evolve. The effectiveness of the systems could also be radically changed by modernization of offensive systems (to include defense suppression systems) and arms reductions.

The Allies. The case for SDI based on security policy, arms control, military requirements, the threat, and technology rationales was never successfully made to the allies in the 1980s. In the political context of the 1990s, barring dramatic reversals in Gorbachev's reforms and in events within Central Europe, that case cannot be made in the Alliance context. But the point may not matter. Despite the yeoman work of spokesmen such as Ambassador Nitze and Assistant Secretary Perle, the allies felt cut out from the outset and have remained so except for the relatively modest number of research contracts let to date with European industries. (See appendix A.)

What the allies needed was an integrative set of insights that captured in a conceptual framework the contributions of defenses to crisis stability and deterrence, to capabilities for denial and defense, and to arms reductions. Even then the burden of proof for change would have remained heavy.

Alliance Strategy. The putative implications of missile defenses on the Alliance's strategic concept has always been too broad and bold for Europeans to accept. The SDI of itself, of course, could never render all nuclear weapons "impotent and obsolete." But Gorbachev's political concessions could render SDI "impotent and obsolete."

Deployed defenses derived from SDI technologies would always be a part, but only a part, of the air defense mission. Other parts included defenses against airplanes and cruise missiles, as well as against the enemy's defense suppression, surveillance, command, control, and communications for battle management and leadership. At the theater level, this approach was exactly what then West German Minister of Defense Woerner had in mind when he pressed NATO for a commitment to antitactical ballistic missile efforts as part of NATO's extended air defenses.

Despite calls from mid-1985 on by several of the allies for a study of the implications of SDI for NATO strategy, no such study could be done without embarrassing consequences for the "vision" and, more important, for the strategy of flexible response. Members of whatever working group the nations commissioned would no doubt have included clever thinkers and drafters capable of papering over chasms of differences in an unclassified executive summary for the press and the public. That was not where the damage would occur. Rather the nations themselves would have to face up to many of the warts on the strategy of flexible response and forward defense, a strategy that Robert McNamara himself may have wished away by 1967.

Not only had the United States and other nations not done their homework for such a study and not only was a study premature, given the state of thinking about strategic defenses and the state of the research efforts, but also the Alliance nations themselves would have had to look in the mirror to find NATO's strategy not the fairest in the land. In sum, there may be a "rough beast" of new strategic truth "slouching toward" Brussels, but it has to do with much more than defenses against ballistic missiles. Most allies knew that and resisted opening the strategy for examination.

The Age of Ballistic Missiles. By the summer of 1988, eight nations had demonstrated their ability to place satellites in space with their own rockets and technologies.[4] Several other nations also possessed intermediate-range and long-range ballistic missiles capable of striking with considerable accuracy from hundreds to thousands of miles from their own territory. This new factor has not yet been fully appreciated in the global military balance. This disturbing development provides yet another warrant for SDI research, testing, and development—the latter activities if confined to fixed, land-based defensive systems could, even under the "narrow" interpretation of the ABM Treaty, proceed all the way to deployment of up to one hundred launchers and one hundred interceptors, with associated radars. For other than fixed land-based systems, testing would have to remain at sub-component level.

Regardless of critics' views about strategic defenses, it is indisputable that ballistic missiles with high-explosive and chemical warheads will be an ever larger part of national arsenals in conflicts ranging from low intensity through large-scale warfare. The city-to-city strikes in the Iran-Iraq war of the spring of 1988 were but the harbinger. Although the United States is not now a prime target for ballistic missile attack by developing third world nations, that may not always be the case in the next few decades. And it is certainly not the situation today for a number of US allies, including Israel.

Some argue with merit that the threat of offensive retaliation is what should continue to deter attack by ballistic missiles. Moreover, there are ways not only to retaliate but also to defend against and to deny enemy capabilities by striking enemy launchers and battle management. For example, tactical air power. However, whatever the truth of these points, the United States and its allies in Western Europe, Japan, Korea, and elsewhere also need now to investigate technologies to learn how best to deny and defend against such attacks. The West has had the decades since the 1944 V–2 ballistic missile attacks on England to think through defense and denial. However, it was primarily through SDI that that additional effort began.

To get a sense of the changed situation, one might ask how the Falklands War might have unfolded if Argentina had had its projected capability with ballistic missiles—or how the ballistic missiles now in the hands of Israel and Iraq might affect any future conflicts in the Middle East. Israel was quick to see the importance of SDI research for theater applications and early on entered a cooperative agreement on research with the United States in 1986.

Wrong Question. President Reagan's challenge was to scientists who had introduced nuclear weapons in the first place, to strategists who had helped base deterrence on the threat of assured destruction, and to arms controllers whose contributions had been to set limits to which the superpowers built their nuclear arsenals. But for the 1972 ABM Treaty, this record did not fill the world with hope.

It was important to challenge the three groups most to blame for nuclear dilemmas and most to praise for the nuclear peace that has prevailed since World War II. The president's question should not have been, "Wouldn't it be better to save lives than to avenge them?" Those are not the only alternatives, and they do not exclude one another. In any event, the president never got a satisfactory answer from any whom he challenged. The correct question was, "Isn't there a better way?"

Promethean Gifts. Edward Teller, Robert Jastrow, James Fletcher, and other scientists favoring SDI rightly cautioned the allies several times not to be too quick to dismiss the kinds of promising technologies and the potential breakthroughs in the research under way. History, they reminded their audiences, was strewn with absurd objections to and fearful skepticism about scientific and technological change; leaders who ought to have known better tried foolishly to hold back waves of discovery, invention, and creativity. Without putting too fine a point on the issue, the character of America in some sense clashed with that of Europe in attitudes toward the possibilities for technology and science to help ensure stability between East and West. America's attempt to make the good better is not an idea that comforts Europeans, but positions can change. Policy alterations in strategy issues, relations with allies, SDI research, and arms control will color the European view.

Strategy Issues Recommendations.
—Undertake a policy review that disciplines and integrates US approaches (including NATO's) to modernization of offensive forces, US positions on strategic arms reductions, the potential contribution of missile defenses to stability, and theater strategies for deterrence. Depending on the results of such a review, engage allies in a military and political assessment of strategic and theater defenses against ballistic missiles of all types and ranges. This assessment should result in a conceptual framework for the way ahead.
—Focus US research efforts on protecting the American homeland, not primarily on defense of land-based strategic ballistic missiles, not even a preferential defense. Do not be apologetic to allies about attempts to reduce the vulnerability of the US homeland and to devalue the worth of ballistic missiles. Cooperate to whatever extent the allies wish in improving theater air defenses, including defenses against ballistic and cruise missiles.
—Recognize the differences between US and European views of vulnerability across the spectrum of possible conflict, especially factors related to geography. As one European defense expert stated the case, flexible response "was created to cope with increasing American *vulnerability* ... [it] is dependent on preserving *a modicum of vulnerability of the Soviet homeland*. This vulnerability will disappear in an unabated defence race, and so will the credibility of extended deterrence."[5] This viewpoint comes close to the heart of the

European concerns about SDI, and therefore it must be understood in fashioning an Alliance consensus.

The Allies' Recommendations.
—Steer clear of creating occasions for the allies to approve or disapprove any policy issues or statements in regard to research and development. Avoid all temptations to make SDI a litmus test of Allied solidarity with the United States. This approach will fail.
—Once a conceptual framework is completed (assuming it underwrites continued research), begin to build a consensus among the allies on a deterrent relationship that retains offensive forces but integrates defense and denial with retaliation as the underpinnings of the strategy. This effort alone might last until the turn of the century. With the Gorbachev-driven changes taking hold, there may be opportunities to build a new strategy based more and more on defensive systems. Tie this effort to defensive concepts prevalent in Europe, East and West. The framework would have to demonstrate the compelling advantages and the tolerable disadvantages to deployment of space-based defenses, or else the commitment will never be accepted by the publics.
—To the extent possible, tighten the policy language used in Alliance circles by senior US officials to describe the drive to devalue the importance of ballistic missiles, to reduce radically the size and destructive potential of the

superpower arsenals, and to have deterrence rely in substantial part on denial and defense.
—Continue the practice of close consultations with the allies on developments in research and on arms control issues related to strategic defenses, both in capitals and at NATO headquarters.

Research Policy Recommendations.
—Continue the current research efforts with the separate management, distinct focus, and funding arrangements of the SDIO. Take some of the pressure off the organization by not insisting on "products" from the research efforts at this point. If the political need to have something to show is overwhelming, then put more energy into translations of the technologies to conventional defense improvements.
—Refrain from pressing hard for additional Allied participation in the SDI research. Let the allies, if they wish, take the role of the demandeur in seeking SDI work.
—Get rid of the idea, currently burdening SDI, that there ever will be a single system to defend the soil of the United States and its allies against attack by ballistic missiles.
—Get across the point that any deployment of strategic, as well as tactical, ballistic missile defenses would be incremental, evolutionary, nondramatic, and measured. At no time could there ever be a complete entity that would not have to be modified to keep up with the enemy's forces, countermeasures, and developments perhaps not even yet thought

of. In any event, nothing will happen soon in deployment, and there is time for consultations with the allies and negotiations with the Soviets.
—Resist deploying terminal defenses of military assets. Unless defenses were deployed in large numbers in excess of what is currently allowed in the ABM Treaty, they would contribute little to the survivability of land-based ballistic missiles, might hasten Soviet offensive and defensive reactions, and would not lessen the vulnerability of the homeland to accidental launch, terrorist attack, or intentional attack in small numbers by the proverbial man or machine out of control.

Arms Control Recommendations.
—Honor the ABM Treaty. Its provisions precluding a territorial defense of any type and its prohibitions of mobile land-based, sea-based, air-based, and space-based ballistic missile defenses are in the US national interest to preserve until such time as technologies for strategic defense of the national soil are available, affordable, and strategically prudent.
—Use the ABM Treaty, when the right time comes in the progress of the technologies, as a vehicle to negotiate the phased, incremental deployment of strategic defenses if that is the national decision.
—Hold on to the idea of convincing the Soviets in Geneva of the merits of defenses in the age of ballistic missiles. They already know the merits well. The thinking to date on both sides has perhaps focused too much on

missiles with nuclear warheads, and not enough on the missiles with chemical and high-explosive warheads now in the hands of a growing number of nations.

—Resist trading SDI away for reductions in strategic arsenals. Deep reductions of even 50 percent in the destructive potential of Soviet land-based ballistic missiles, the merit of which stands on its own, would not change the nature of deterrence and the threat of annihilation underlying it. That said, SDI can be used to help establish a future regime and a new balance much less reliant on ballistic missiles. What needs to be thought through and negotiated, once the research proves out, is a series of transitional phases to lessen any incentives to strike first in a crisis. The Soviets may be more ready to agree to defensive arrangements than ever before.

APPENDICES

APPENDIX A
STATUS OF ALLIED CONTRACTS

Country	Number of Contracts	$ Values (M)
United Kingdom	103	76.95
West Germany	33	65.93
Israel	16	184.50 *
Italy	25	14.15
Japan	11	2.20
France	10	12.70
Canada	16	2.51
Belgium	1	.09
Denmark	1	.03
The Netherlands	2	12.04 **
Total	218	$371.1

* Includes $31.6 million contribution by Israel
** Includes $7.0 million contribution by The Netherlands
Source: SDIO

APPENDIX B
SDI FUNDING LEVELS

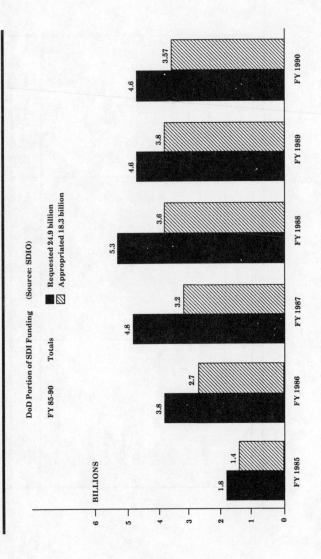

NOTES

1. TO AVENGE OR TO DEFEND?

1. See Robert W. Helm, "The Strategic Defense Initiative: Its Genesis and Transformations," in *The Strategic Defense Initiative: An International Perspective*, ed. C. James Haug (Boulder, Colorado: Social Science Monographs, 1987), pp. 1-13. At the time of the drafting of the president's March 23, 1983, speech launching SDI, Mr. Helm was the director of the Defense Programs and National Security Telecommunications Policy Division of the National Security Council. Mr. Helm provides "firsthand" assessments of the "sense of strategic frustration in US policy" which motivated the president in announcing SDI. See also Michael Mandelbaum and Strobe Talbott, *Reagan & Gorbachev* (New York: Vintage Books, 1987) and Strobe Talbott, *The Master of the Game* (New York: Vintage Books, 1989), pp. 185-87.

2. For examination of how SDI was treated in the communiques of the ministerial meetings of the Nuclear Planning Group, the Defense Planning Committee, and the North Atlantic Council—better known to many as the NPG, the DPC, and the NAC, respectively—see *NATO Final Communiques 1981-1985*, *NATO Communiques 1986*, *NATO Communiques 1987*, and *NATO Communiques 1988*—all published by NATO Information Service, Brussels, Belgium. By far and away the strongest communique language supporting SDI research, as "in NATO's security interest" came at the Luxembourg NPG, March 26 and 27, 1985. Ministers declared the program "prudent in the light of ... Soviet activities."

3. For discussion of the potential contributions of the "defenses" to "offenses," see Lawrence Freedman, *Strategic Defence in the Nuclear Age*, Adelphi Papers, no. 224 (London: International Institute of Strategic Studies, 1987), pp. 4-20 and passim; see also Harald Muller, *Strategic Defences: The End of the Alliance Strategy* (Brussels: Centre for European Policy Studies, 1987). The very idea that certain weapons belong to the offense and certain others strictly to the defense is a flawed concept to this author.

4. See Lawrence Freedman, *Strategic Defence in the Nuclear Age*, p. 17, for a discussion of the terms *deterrence*, *defence*, *offense*,

and *compellance*, the latter is Thomas Schelling's term to distinguish between *inducing inaction* (deterrence) and *making someone perform* (compellance).

5. The administration began to use annual editions of *Soviet Military Power* as a way to publicize Soviet legal and illegal efforts to deploy extensive civil defenses, air defenses, and ballistic missile defenses. Following the issuance of NSDD 172, the State Department published its Special Report No. 129 in June 1985. The Departments of Defense and State also jointly issued and widely distributed a glossy booklet entitled *Soviet Strategic Defense Programs* in October 1985.

6. Although not signatories to the ABM Treaty of 1972 and its Protocol of 1974, European nations have retained a proprietary interest in it. For example, one of the strong recommendations of the Muller analysis (see note 3 above) is "Preserve Soviet Vulnerability: save the ABM Treaty." Even though Muller acknowledges the possibility of a different agreement serving strategic stability, he states that "preserving flexible response means maintaining a modicum of Soviet vulnerability: an ABM Treaty banning area defences ... would serve this purpose well" (p. 33). Prime Minister Thatcher included the ABM Treaty in the points she worked out with President Reagan in December of 1984. Moreover, both France and the United Kingdom want to keep the ABM Treaty in place and prevent strategic defenses in order to preserve the credibility and independence of their own nuclear arsenals.

7. In a 1985 BBC interview, Hoffman said this of President Reagan's goals of rendering nuclear weapons "impotent" and "obsolete": "the President used very general terms in that speech; and some of them clearly have to be viewed as Presidential rhetoric." See Michael Charlton, *The Star Wars History: From Deterrence to Defence: The American Strategy Debate* (London: BBC Publications, 1986), p. 111.

8. For a review of survivability issues, see Brent Scowcroft, John Deutch, and R. James Woolsey, "Verify But Survive," *Washington Post,* June 14, 1988, p. A21.

9. See *Discriminate Deterrence:A Report of the Commission on Integrated Long-Term Strategy*, Co-Chairs Fred C. Ikle and Albert Wohlstetter, January 1988. In this document, Ikle and Wohlstetter explore many themes about non-nuclear, effective, highly accurate conventional weapons, as well as the attempts to make

nuclear weapons more discriminate and therefore supposedly more credible. In the same document, questions are raised about the value of NATO's deterrent strategy for the long term.

10. See the White House pamphlet entitled *The President's Strategic Defense Initiative*, January 1985.

11. A White House issue brief on the Reykjavik summit, dated October 16, 1986, developed many of these concepts.

12. In sum, the treaty (as amended in 1974) allowed deployment of up to 100 launchers, 100 missiles, and associated radars to protect one site in the homeland, either the capital (which the Soviets selected as expected) or military assets (the United States deployed the Sentinel/Safeguard system for a brief period in 1975/76). Prohibited were the development and deployment of the ABM components (missile interceptors, launchers, and radars) of an air-based, sea-based, mobile land-based, or space-based ballistic missile defense of the territory.

However, Agreed Statement D of the treaty appears to prohibit explicitly only the deployment, not the development and testing, of ABM components based on other physical principles, that is, on principles other than those on which ABM systems were based at the time of the treaty ratification in 1972. In the view of many in the Reagan administration (including Secretaries Shultz and Weinberger as well as Ambassador Nitze), by this statement the United States and the Soviet Union allowed the development and testing of ABM components (no matter how eventually to be deployed) based on other physical principles. However, deployment of anything other than the 100 fixed, land-based systems allowed in the 1974 amendment to the treaty would still be prohibited. With the "broad" interpretation, or what the Reagan administration termed the "legally correct interpretation" of the treaty, the SDIO would have been able to perform tests of and proceed to full scale development of ballistic missile defense components that could eventually be deployed in ways prohibited by the ABM Treaty. Only the deployment itself would be prohibited.

Most European leaders, as well as many powerful members of the Congress like Senator Nunn, balked at the "broad" interpretation and looked at it as a way for SDI zealots in the administration and in the scientific community to go beyond current restrictions on tests and development that had guided the SDI research from the outset of the program. Moreover, congressional critics of the reinterpretation saw in it a potentially serious constitutional issue between the Executive and the Congress over the ratification of treaties. Congress rejected the concept that after the ratification process the Executive could develop interpretations not explicated and commonly understood at the time of the ratification hearings.

13. See the discussion of the ABM Treaty in Jacquelyn K. Davis and Robert L. Pfaltzgraff, Jr., *Strategic Defense and Extended Deterrence: A New Transatlantic Debate*, National Security Paper, no. 4 (Cambridge, Massachusetts: Institute for Foreign Policy Analysis, Inc., February 1986), pp. 2 and 3. According to Davis and Pfaltzgraff, "Nowhere has the attachment to the mutual assured destruction deterrence concept been stronger than in Western Europe. Central to this paradigm is the ABM Treaty, which West German Foreign Minister Hans-Dietrich Genscher has described as the 'Magna Carta' of arms control." British Foreign Secretary Geoffrey Howe is quoted saying that the ABM Treaty's effect was "to enhance the strategy of nuclear deterrence through the clear recognition of mutual vulnerability."

See also William J. Durch, *The Future of the ABM Treaty*, Adelphi Papers, no. 223 (London: The International Institute of Strategic Studies, 1987).

14. See Ambassador Nitze's Alastair Buchanan Memorial Lecture to the International Institute for Strategic Studies, London, March 28, 1985. The speech is included in USIS Wireless File No. 60, March 29, 1985. See also Michael Charlton, *The Star Wars History: From Deterrence to Defence: The American Strategic Debate* (London: BBC Publications, 1986), p. 55. In the Charlton text, Dr. Kissinger claims that the Nitze criteria are "in fact unfulfillable" and therefore Nitze is really working for conciliation with the Soviet Union. Moreover, Secretary Weinberger at every opportunity underlined the idea that "cost effectiveness" was more than an economic concept. He also for a time attempted to substitute the idea of "affordability," arguing that the nation could afford strategic defenses if they were highly effective.

15. See Michael Charlton, *The Star Wars History*, pp. 26-27.

16. Ibid., p. 27.

17. Benjamin Lambeth and Kevin Lewis, "The Kremlin and SDI," *Foreign Affairs* 66 (Spring 1988): 755-70.

18. Ibid., p. 759.

19. White House pamphlet, p. 2. See also Charles Krauthammer, "Reykjavik and the End of Days: A Triumph of

American Unilateralism," *New Republic*, November 17, 1986, pp. 22-25, 28.
  20. White House pamphlet, p. i.
  21. Michael Charlton, *The Star Wars History*, p. 61.
  22. Jeane Kirkpatrick, "Dukakis' Dangerous Ideas," *Washington Post*, June 20, 1988, p. A13.
  23. Charles Krauthammer, "Reykjavik and the End of Days," pp. 25, 28.
  24. Quoted in Harald Muller, *Strategic Defences*, p. 11.
  25. See the excellent discussion of the evolution of strategic objectives in Richard I. Brody, *Strategic Defences in NATO Strategy*, Adelphi Papers, no. 225 (London: International Institute of Strategic Studies, 1987).

2. EUROPEAN CANDLING OF SDI

  1. The term "European" is obviously an oversimplification, given the variety of nations, cultures, viewpoints, and so forth among the fifteen partners with the United States in the North Atlantic Treaty Organization. Canada and Iceland, as well as possibly Turkey, are not Europe's own. General de Gaulle believed the Anglo-Saxons had left Europe for England. That said, the author begs the indulgence of readers in allowing the term "European" to denote, the Olympics in Calgary notwithstanding, non-US members of the Alliance. In chapter 3, the particular nuances of reactions to SDI by individual nations have an airing.
  2. This complaint has had many voices in Europe. See especially Harald Muller, *Strategic Defences: The End of the Alliance Strategy* (Brussels: Centre for European Policy Studies, 1987), p. 6 and passim. Muller's monograph provides a good but overly negative catalogue and explication of European perspectives on SDI.
  3. Pierre Lellouche, *L'IDS et L'Alliance atlantique: options politiques et strategiques* (Paris: Institut Francais des Relations Internationales, 1986),pp. 165, 157-73. See also Lellouche, "SDI and the Atlantic Alliance," *SAIS Review*, Summer/Autumn, 1985.
  4. See Michael Charlton, *The Star Wars History: From Deterrence to Defence: The American Strategic Debate* (London: BBC Publications, 1986), p. 17.
  5. Quoted in Michael Charlton, p. 26.
  6. Ibid., pp. 26-27.

7. For example, Chancellor Helmut Kohl in an April 1985 speech to the Bundestag helped set the pace in European nations for support of the research but for caution in regard to strategy implications of deployed systems. Those expressing rather complete support have included General Pierre Gallois (French Air Force, Ret.); Air Vice-Marshal Stewart W. B. Menaul (Former Chief of Staff, RAF Bomber Command), a principal advocate of the European Defense Initiative (EDI); and Lord Chalfont. *The Economist* also supported SDI quite early; see the editorial in the August 3, 1985, issue.

8. Fred S. Hoffman (Study Director), *Ballistic Missile Defenses and U.S. National Security, Summary Report*, prepared for the *Future Security Strategy Study*, October 1983. See also James C. Fletcher, *The Strategic Defense Initiative, Defensive Technologies Study*, a Department of Defense summary of the classified Fletcher study published April 1984.

9. Included as a transmittal note to the April 1984 unclassified summary of the James C. Fletcher study, *The Strategic Defense Initiative, Defensive Technologies Study*.

10. Ibid., p. 13. Text included in Steven E. Miller and Stephen Van Evera, eds., *The Star Wars Controversy* (Princeton: Princeton University Press, 1986), p. 313.

11. Included in Steven E. Miller and Stephen Van Evera, *The Star Wars Controvery*, p.279.

12. The principal US negotiators briefed allies at NATO headquarters at the beginning and at the end of each round of arms control negotiations with the Soviets.

13. See Larry Pressler, *Star Wars: The Strategic Defense Initiative Debates in Congress* (New York: Preager, 1986), p. 148.

14. For a well-informed discussion of early expectations for participation, see Lothar Ibrugger (Rapporteur), *General Report on Strategic Defence: Technology Issues*, Scientific and Technical Committee, North Atlantic Assembly, November 1986.

15. Lord Carrington discussed SDI in a number of public speeches in this period.

16. Arnold Kanter, "Thinking About the Strategic Defense Initiative: An Alliance Perspective," *International Affairs* 61 (Summer 1985): 449-64. Kanter thought that SDI would launch a long-overdue debate on security in the nuclear age and cast that "in sharper terms."

17. See Abraham D. Sofaer (legal adviser to the State Department), statement before the subcommittee on arms

control, international security, and science of the House Committee on Foreign Affairs, on October 22, 1985. Sofaer stated that his "study of the [ABM] Treaty led [him] to conclude that its language is ambiguous and can more reasonably be read to support a broader interpretation." The previous May, Ambassador Nitze (special advisor to the president and secretary of state for arms control), himself one of the ABM Treaty negotiators, said the treaty "was intended to be adaptable to new circumstances"; his speech was delivered at the commencement ceremony for the Johns Hopkins School of Advanced International Studies on May 30, 1985. See also Abraham D. Sofaer, "The ABM Treaty: Legal Analysis in the Political Cauldron," *Washington Quarterly*, Autumn 1987, pp. 59-75. And Sam Nunn, "The ABM Reinterpretation Issue," *Washington Quarterly* 10 (Autumn 1987): 45-57.

18. On this side of the Atlantic, the niceties of the Carrington "firebreak"—a term sometimes also used of the break between theater nuclear weapons and central strategic systems—were not so important. After all, the initial Hoffman and Fletcher studies spoke of goals for intermediate systems that could be ready in the near term; there was little to no doubt about how any deployment decision would come out.

19. David M. Abshire, *NATO on the Move*, The Alliance Papers, no. 6 (Washington, DC: The Atlantic Council, September 1985).

20. In January 1985, for example, the White House published and widely distributed a pamphlet entitled "The President's Strategic Defense Initiative." Needless to say, the pamphlet did not resolve the basic tension and contrarieties between population defense and defense of military assets, nor between the near-term pressure for deployment (in the Hoffman strain) and the program for strategic defenses in the long term (in the Fletcher strain). In October 1985, the State and Defense Departments jointly issued a report on "Soviet Strategic Defense Programs."

21. At the US Mission to NATO, Ambassador Abshire established a "Truth Squad" to respond in one to two days to articles in the European press that distorted SDI in any way. This work caused dark rings under many eyes.

22. See especially Lawrence Freedman, *Strategic Defence in the Nuclear Age*, Adelphi Papers, no. 224 (London: The International Institute for Strategic Studies, 1987). See also Ivo H.

Daalder, *NATO Strategy and Ballistic Missile Defence*, Adelphi Papers, no. 233 (London: The International Institute for Strategic Studies, Winter 1988).

23. In this speech, Ambassador Nitze said, "We are even now looking forward to a period of transition to a more stable world.... A world free of nuclear arms is an utlimate objective...."

24. See discussion in Lothar Ibrugger, *General Report on Strategic Defence; Technology Issues*, pp. 22-29. See also appendix A for a chart on SDI contracts let to European firms. On July 21, 1987, Japan and the United States signed an agreement on participation in SDI research.

25. Quoted in Lothar Ibrugger, *General Report on Strategic Defence: Technology Issues*, p. 28.

26. James Woolsey, "Memo for: SDI Supporters & Critics: Recommendation: Try Collective Security," *Armed Forces Journal*, September 1985, p. 98.

27. Manfred Woerner, "A Missile Defense For NATO Europe," *Strategic Review* 14 (Winter 1986): 13-20. See also "German Minister Discusses NATO's Defense Options," *Aviation Week & Space Technology*, November 17, 1986, pp. 77, 79. In addition, General Rogers, then SACEUR, was also calling for work on a European defense—see Bob Furlong and Macha Levinson, "SACEUR Calls for Research on a European ABM System," *International Defense Review*, February 1986, p. 149.

28. Manfred Woerner publicly discounted the notion of an EDI in the spring of 1986 at a press conference on the margins of the Nuclear Planning Group ministerial at Wurtzburg. However, see C. James Haug, *The Strategic Defense Initiative: An International Perspective* (Boulder, Colorado: Social Science Monographs, 1987. In the latter work (p. 76), Air Vice Marshal Menaul briefly discusses his work, along with General Gallois, on EDI. Menaul insists there is nothing new with the idea of trying to defend against ballistic missiles; in fact, it began not with SDI, but "when the first ballistic missile fell on London."

29. As a result of the Woerner interest in ATBM, NATO began in earnest to examine the threat and potential countermeasures to tactical ballistic missile attack, including active, passive, and counterbattery measures, as well as tactical air power. This work has been reported on over the past several years in North Atlantic Council (NAC) and Defense Planning Committee communiques. The NATO Air Defence Committee (NADC)

and the Advisory Group on Aerospace Research and Development (AGARD), in conjunction with SHAPE's efforts, have done most of the Alliance's in-house work.

30. See Abraham D. Sofaer, "The ABM Treaty and the Strategic Defense Initiative," *Harvard Law Review* 99 (June 1986): 1985.

31. *Soviet Strategic Defense Programs*, a publication released by the Departments of Defense and State, October 1985. In addition, see annual publications of DOD's *Soviet Military Power*.

32. Glenn A. Kent and Randall J. DeValk, *Strategic Defenses and the Transition to Assured Survival* (Santa Monica, California: Rand Corporation, October 1986), p. viii.

33. Caspar Weinberger, speech to the Commonwealth Club, April 27, 1987. Included in the "USIS European Wireless File," no. 80, April 28, 1987.

34. Ibid.

35. Harald Muller, *Strategic Defences*, p. 4.

36. See Albert Wohlstetter, "Swords Without Shields," *National Interest*, 5 (Summer 1987): 31-57.

37. For example, see Keith B. Payne, *Strategic Defense: "Star Wars" in Perspective* (Lanham, Maryland: Hamilton Press, 1986), as well as Colin Gray, "Strategic Defences: A Case for Strategic Defence, *Survival* 27 (March/April 1985): 50-55.

38. For a survey of European thinking about the Reykjavik summit's potential effects on the Alliance, see Karl Kaiser, "The NATO Strategy Debate After Reykjavik," *NATO Review*, December 1986, pp. 6-13. As Harald Muller has put it (p. 14), "The sweeping proposals discussed at Reykjavik ... indicate that extended deterrence and flexible response were not high priorities in the minds of those negotiating." In a report to Congress, in late January 1988 the JCS themselves concluded that elimination of all US and Soviet ballistic missiles, as first suggested by President Reagan at the 1986 summit meeting, could be costly and dangerous.

39. Lawrence Freedman, *Strategic Defense in the Nuclear Age*, pp. 8–10.

40. Arnold Kanter, "Thinking About the Strategic Defense Initiative," p. 452. See also Jeffrey Simon, ed., *Security Implications of SDI* (Washington, DC: US Government Printing Office, 1990).

41. Quoted in Glyn Ford, "The Dangers of SDI," *The European* 1 (September/October 1987): 15.

42. Ibid., pp. 14-15.
43. Harald Muller, *Strategic Defences*, p. 15.
44. Ibid., p. 32.
45. Paul Nitze, speech and article for the Institute for Theology and Peace, Bonn, West Germany, December 9, 1985. The article was carried in the USIS Wireless File, no. 236, December 10, 1985.

3. SDI AT EUROPEAN CAPITALS AND NATO

1. Senator Larry Pressler, *Star Wars: The Strategic Defense Initiative Debates in Congress* (New York: Praeger Special Studies, 1986), pp. 149-50. Senator Pressler provides a good summary of European views on SDI (pp. 144-55), including a discussion of British Foreign Secretary Sir Geoffrey Howe's remarks of March 15, 1985, expressing reservations about SDI.

2. By the late May 1985 Defense Planning Committee and the early June North Atlantic Council ministerial meetings, all references to strategic defenses were indirect (at best) expressions of support for "efforts in all three areas of negotiations" at Geneva.

3. By the December 1985 ministerials, the allies "expressed strong support for the United States stance concerning intermediate-range strategic and defence and space systems."

In the May 1986 DPC ministerial communique, Greece, Denmark, and Norway reserved their positions on "defence and space systems." However at the following fall Nuclear Planning Group (NPG) ministerial meeting at Gleneagles, Scotland, the following endorsement was crafted late at night: "We strongly support the United States exploration of space and defence systems, as is permitted by the ABM Treaty." Needless to say, many had their private meanings attached to the comma before the words "as is permitted."

The December 1986 DPC ministerial communique had similar language. This was the last time the United States sought endorsing language on SDI in communiques.

4. The nations requesting footnotes because of wording on SDI included Greece, Denmark, and Norway (on two occasions). The term *footnote nation* began to appear in discussions outside and inside NATO proper, not only in the press but also occasionally in some of the European parliaments during question time with ministers. Norway in particular found the term

demeaning and worked diligently to find wording that would avoid any reason to footnote the communiques. Norway decidedly did not want the "footnote nation" status that Denmark had achieved.

5. The US negotiators for the Geneva talks briefed the allies in Brussels, usually before and after each round of negotiations with the Soviets. Since SDI was one of the most contentious of the issues, the sessions in Brussels would include discussion of SDI in the arms control context of the ABM Treaty. Moreover, from time to time, the secretary of state or the secretary of defense would also address SDI with the allies, particularly at ministerial sessions several times per year. In addition, there were frequent exchanges on SDI between the NATO ambassadors, and occasional public and private sessions with senior US officials such as SDIO Director General Abrahamson and Ambassadors Nitze and Rowny, as well as Ambassador Abshire and Assistant Secretaries of Defense and State Perle and Ridgway.

6. In the January 1988 *Report of The Commission On Integrated Long-Term Strategy* entitled *Discriminate Deterrence*, on page 27 co-chairs Dr. Fred Ikle and Dr. Albert Wohlstetter assert that "NATO plainly needs a coherent strategy that will be viable for the long haul." Evidently, NATO did not have such a strategy in the eyes of the commission members in 1988 even before the revolutionary events of the next two years.

7. Ambassador Paul Nitze, address before the North Atlantic Assembly, San Francisco, October 15, 1985. The address was entitled "SDI: Its Nature and Rationale."

8. Lawrence Freedman, *Strategic Defence in the Nuclear Age*, Adelphi Papers, no. 224 (London: International Institute for Strategic Studies, Autumn 1987).

9. Charles L. Glasser, "Why Even Good Defenses May Be Bad," in *The Star Wars Controversy*, eds. Steven E. Miller and Stephen Van Evera (Princeton University Press, 1986), p. 56.

10. Congress of the United States, Office of Technology Assessment, *Ballistic Missile Defense Technology* (Washington, DC: US Government Printing Office, 1985). On page 33, one of the important findings of this study was that unless the imbalance between the advantages of offensive strategic technologies over defensive technologies were redressed, "strategic defenses might be plausible for limited purposes, such as defense of ICBM silos or complication of enemy attack plans, but not for the more

ambitious goal of assuring the survival of U.S. society." In short, "assured survival of the U.S. population appears impossible to achieve if the Soviets are determined to deny it."

In this vein, the Aspen Strategy Group also studied and reported, in October 1986, on the strategy, technology, and arms control implications of the SDI program. See *The Strategic Defense Initiative and American Security* (Lanham, Maryland: The University Press of America, 1986). On page v, the Aspen report took a position similar to that of the OTA; namely, although the "President's vision is both clear and desirable, it is not realistic within any operative time frame. We see virtually no prospect of building a significant and effective population shield against a responsive enemy inside this century, and there is great uncertainty about the long term."

11. Glenn A. Kent and Randall J. DeValk, *Strategic Defenses and the Transition to Assured Survival* (Santa Monica, California: Rand Corporation, 1986). Among Rand's substantial work on SDI such as that of James Thomson, General Kent's approach is clearly the most important for SDI advocates and other proponents of defensive strategies to understand, especially for any discussion on potential transitions.

12. A number of SDI advocates such as Robert Jastrow, Fred Hoffman, George Keyworth, and particularly Edward Teller debated scientific SDI issues in public and private with leaders of the Union of Concerned Scientists such as Richard Garwin and the Nobel Prize winner Hans Bethe. In Europe, Hans Ruehle (head of the Planning Division of the West German Ministry of Defense) responded forcefully to rebut the "scientific" criticism of SDI. See "Toward the Limit of Technology," *Der Spiegel*, no. 48, 1985.

13. Robert Jastrow (former director of the Goddard Space Flight Center and currently a professor at Dartmouth College) gave an interview to the US Information Agency in December 1985; the interview was carried in the USIS European Wireless File, no. 236 (December 10, 1985). As he also did in his later writings on SDI, Jastrow presented a strong case in that interview for proceeding with SDI. He also challenged the Union of Concerned Scientists directly, commenting that it is "an organization mostly of non-scientists. Anybody with $15.00 can join. They have put out a statement ... which was signed by 54 Nobel laureates from the United States ... saying, borrowing from the Soviet rhetoric, that they are against the militarization of space.

It's interesting to note that of the 54 Nobel laureates, only one, Hans Bethe, has any contact with defense matters, any experience in missile defense whatsoever."

14. Karl von Clausewitz, *On War*, translated and edited by Peter Paret and Michael Howard (Princeton, New Jersey: Princeton University Press, 1976), p. 357: "The defensive form of war is not a simple shield, but a shield made up of many blows."

15. Michael J. Deane and Ilana Kass, "Why Strategic Defense But Not Defensive Strategy," *Signal* 42 (November 1987): 52-58. Soviet "ideology closely intertwines the ideas of systemic defense and systemic offense as mutually reinforcing requirements."

16. For useful perspectives on Allied participation in ballistic missile defense research, see Lothar Ibrugger, *Strategic Defence: Technology Issues*, North Atlantic Assembly, November 1986. See also David S. Yost, "Ballistic Missile Defense and the Atlantic Alliance," in *The Star Wars Controversy*, eds. Steven E. Miller and Stephen Van Evera (Princeton, New Jersey: Princeton University Press, 1986), pp. 131-62. Yost's work overall is excellent on European and American views on BMD. See especially David S. Yost, "European Anxieties About Ballistic Missile Defense," *Washington Quarterly* 7 (Fall 1984): 112-28.

17. See Keith B. Payne, *Strategic Defense: "Star Wars" in Perspective* (Maryland: Hamilton Press, 1986), p. 200, and Senator Larry Pressler, *Star Wars: The Strategic Defense Initiative Debates in Congress*, pp. 153-55.

18. For a discussion of French attitudes, see Larry Pressler, *Star Wars: The Strategic Defense Initiative Debates in Congress*, pp. 151-55.

19. Lothar Ibrugger, *Strategic Defence: Technology Issues*, p. 28, quotes Chirac as saying, "France cannot afford not to be associated with this great research programme."

20. Keith B. Payne, *Strategic Defense*, p. 201. See Larry Pressler, *Star Wars*, p. 154. Pressler also quotes West German Foreign Minister Hans Dietrich Genscher in late May 1985, asserting that Europe "must not drop to the level of subcontractor and supplier" in the SDI program.

21. Lothar Ibrugger, *Strategic Defence: Technology Issues*, p. 23.

22. See David A. Brown, "European Industry Begins To Seek U.S. SDI Contracts," *Aviation Week & Space Technology*,

December 16, 1985, pp. 12-15. See also "Britain Signs MOU to Participate in SDI," *Aviation Week & Space Technology*, December 16, 1985, p. 12.

23. "Britain Signs MOU," p. 12.

24. Senator Glenn both in 1986 and again in 1987 offered an amendment that would have given US industry a price advantage over foreign companies bidding for SDI contracts. While there was much superficially attractive to the amendment in the United States, it would have had a chilling effect in Europe on armaments cooperation between the United States and Alliance nations. Once again, the inconstancy of the American political system was clear to Europeans—no decisions in the US Government ever appeared to be final. Ambassador Abshire led the charge from Europe against the amendment.

25. Keith B. Payne, *Strategic Defense*, p. 193.

26. This quote and the one that follows concerning the four points are taken from Mrs. Thatcher's press statement of December 22, 1984, issued after her meetings with the president.

27. Larry Pressler, *Star Wars*, pp. 149-50. See also Sir Geoffrey Howe, "Defence and Security in the Nuclear Age," speech delivered March 15, 1985, to the Royal United Services Institute, London, England.

28. *The Economist*, in a strongly-worded editorial of August 3, 1985, page 11, supported SDI, arguing that "it can work, and it can help." The editorial stated a particular case for SDI: "If America had this protection and Russia did not, it would also make the nuclear umbrella the Americans hold over Europe and Japan a lot less threadbare than it is now, when America is naked to missiles."

29. See Menaul's argumentation in *The Strategic Defense Initiative: An International Perspective*, ed. C. James Haug (Boulder, Colorado: Social Science Monographs, 1987), pp. 65-100. This text records the proceedings of a conference on SDI sponsored by the Center for International Security and Strategic Studies, Mississippi State University.

30. "Howe's UDI from SDI," *Times* (London), March 18, 1985, p. 13. Quoted in Keith B. Payne, *Strategic Defense*, p. 199.

31. C. James Haug, *The Strategic Defense Initiative*, p. 76. According to the US Air Force, Office of Air Force History, the first V-2 rocket fell on London on September 8, 1944, not the 16th of April as printed in Haug. It was launched from Holland.

32. Ibid., p.79.
33. See the discussion of Reykjavik in chapter 2.
34. George Younger, "Europe or America, A False Dilemma," a Paper delivered to the Wehrkunde Conference in Munich in early February 1987.
35. See also Keith B. Payne, *Strategic Defense*, pp. 200-201, and Larry Pressler, *Star Wars*, pp. 151-52.
36. See Maxwell Taylor, *The Uncertain Trumpet* (New York: Harper and Brothers, 1959), pp. 158-62. These pasages discuss the "kinds and quantities" of forces required to support a National Military Program of Flexible Response [page 162]."
37. Michael Charlton, *The Star Wars History: From Deterrence to Defence* (London: BBC Publications, 1986), pp. 1-28 and passim.
38. Ibid., pp. 17, 19, and 23
39. Ibid., p. 130.
40. Ibid., p. 18.
41. Larry Pressler, *Star Wars*, p. 151.
42. Michael Charlton, *The Star Wars History*, p. 116.
43. Larry Pressler, *Star Wars*, p. 152.
44. Keith B. Payne, *Strategic Defense*, pp. 200-201.
45. *Eighth German-American Roundtable on NATO: Strategic Defense, NATO Modernization, and East-West Relations* (Cambridge, Massachusetts: Institute for Foreign Policy Analysis, 1986), passim. This December 1985 conference was jointly sponsored by the institute and the Konrad-Adenauer-Stiftung.
46. Some feel Kohl *let* Genscher take the lead as a means of deflecting attention from *his own* misgivings. Moreover, SDI opposition in the Bundestag went far beyond the SPD leadership, including most of the SPD and a significant number in Kohl's own CDU (though their reasons differed). By 1985, Schmidt was a totally spent force in the SPD.
47. Hans Ruehle, "Toward the Limit of Technology," *Der Spiegel*, no. 48, 1985.
48. Larry Pressler, *Star Wars*, pp. 146-48.
49. Keith B. Payne, *Strategic Defense*, pp. 194-97.
50. White House pamphlet, *The President's Strategic Defense Initiative*, January 1985.
51. Ibid., p. 7.
52. In recent years, Robert S. McNamara himself has cast considerable doubt on what he thought of NATO's strategic concept by 1967, after five years of convincing the allies to agree

to it (with the exception of France). The limited nuclear options in the European theater made little sense to him.

53. Quoted in Elizabeth Pond, "Defense Minister Defends 'Star Wars'," *Christian Science Monitor*, April 12, 1986, p. 7.

54. Christoph Bertram, "Strategic Defence in Europe," *NATO's Sixteen Nations*, June 1986, pp. 28-30. See also excerpts from a Weinberger-Woerner Press Conference, Giessen, West Germany, on February 12, 1985. USIS Wireless File, no. 29, February 13, 1985.

55. White House issue brief, October 16, 1986.

4. IMPLICATIONS FOR NATO'S STRATEGY

1. Strobe Talbott, "Inside Moves," *Time*, May 30, 1988, p. 35-36.

2. For example, R. Jeffrey Smith, in a *Washinton Post* article, July 16, 1988, page A21, entitled "U.S. ABM Treaty Complaint Imperils START, Soviet Says," reported that the "administration is sharply divided over a proposal by the civilian leadership of the Defense Department under which the United States could withdraw from all or part of the ABM Treaty on grounds that [Krasnoyarsk] is a 'material breach' of the accord." The article included statements from Ambassador Nitze and the JCS representative arguing against such announcement of an intention to withdraw. See also "U.S. Rejects Soviet Bid to Link Radar, ABM Pact," *Washington Post*, July 21, 1988, A11. It was not until November 1989 that the Soviet Union admitted that the radar construction was a breach of the treaty—and a stupid one at that. They pledged to dismantle it; the West awaits.

3. Quoted in Michael Charlton, *The Star Wars History* (London: BBC Publications, 1986), p. 14. Odom also said, "To take the Western deterrence view of 'assured destruction' is really to say that we surrender primacy over our choice about matters of war to technical gimmickry ... [we just] sit back and let somebody else make the choice about whether deterrence will fail or not.... From time to time I have described it as Ptolemy's view of the solar system. When the stars do not behave the way you want them to, you create one more little epicycle to account for this erratic behavior by the Soviets. I think the Soviets have a Copernican view of the solar system."

4. Ibid., p. 15.

5. Ibid., p. 17.
6. See chapter 3 for discussion and elaboration of this thinking by Harald Muller.
7. See especially chapter 2 of this book.
8. See Lawrence Freedman, *The Evolution of Nuclear Strategy* (New York: St. Martin's Press, 1981), pp. 305-7, 397-400.
9. Harald Muller, *Strategic Defences: The End of the Alliance Strategy* (Brussels, Belgium: Centre for European Policy Studies, 1987), p. 10.
10. Ibid., p. 11.
11. Quoted in Harald Muller, p. 11; see G. Fricaud-Chagnaud, "Active Deterrence, Active Solidarity," *Defence Analysis* 2 (March 1986): 4 ff.
12. Pierre M. Gallois, "US Strategy and the Defense of Europe," *Orbis* 7 (Summer 1963), quoted in Lawrence Freedman, *The Evolution of Nuclear Strategy*, p. 301.
13. Ibid., p. 303.
14. Samuel P. Huntington, "Conventional Deterrence and Conventional Retaliation in Europe," *International Security* 8 (Winter 1983-84): 32-56.
15. *NATO: Facts and Figures* (Brussels: NATO Information Service, 1976), p. 33. "The Ministers adopted the firm force goals proposed by the Temporary Council Committee, a total of 50 divisions, 4,000 aircraft and strong naval forces by the end of 1952, as well as their provisional estimates for 1953 and 1954." The estimates called for more than ninety divisions.
16. Fred S. Hoffman, *Ballistic Missile Defenses and U.S. National Security: Summary Report*, October 1983, p. 1. Prepared for the Future Security Strategy Study.
17. Ibid., p. 1. See also Fred C. Ikle and Albert Wohlstetter (co-chairmen), *Discriminate Deterrence: Report of the Commission on Integrated Long-Term Strategy* (Washington, DC: US Government Printing Office, January 1988), p. 2: "Both our conventional and nuclear posture should be based on a mix of offensive and defensive systems. To help deter nuclear attack and to make it safer to reduce offensive arms we need strategic defense. To deter or respond to conventional aggression we need a capability for conventional counter-offensive operations deep into enemy territory."
18. Samuel P. Huntington, "Conventional Deterrence and Conventional Retaliation in Europe," p. 34.

19. Lawrence Freedman, *The Evolution of Nuclear Strategy*, p. 396.
20. Ibid.
21. See Strobe Talbott, "Inside Moves," p. 32.
22. Lawrence Freedman, *The Evolution of Nuclear Strategy*, pp. 396-97.
23. Ibid., p. 397.
24. Fred C. Ikle and Albert Wohlstetter, *Discriminate Deterrence*, p. 36.
25. Lawrence Freedman, *The Evolution of Nuclear Strategy*, p. 399. See also the Roman Catholic bishops' report on their earlier pastoral letter entitled "The Challenge of Peace." The bishops, with Father Brian Hehir as principal drafter, sought to make clear that not just the asserted intentions of SDI but also SDI's potential and actual consequences should be determinants in any judgment about whether deterrence based on strategic defenses would be any more "moral" than deterrence based on the threat of offensive nuclear systems.
26. See also Keith B. Payne, *Strategic Defense: 'Star Wars' in Perspective*, p. 96. Payne's discussion of the implications of SDI for strategic stability is a good survey of views. The Scowcroft Commission examining strategic modernization in 1982 defined strategic stability in its more-or-less accepted meaning as the "condition which exists when no strategic power believes it can significantly improve its situation by attacking first in a crisis or when it does not feel compelled to launch its strategic weapons in order to avoid losing them."
27. Glenn A. Kent and Randall J. DeValk, *Strategic Defenses and the Transition to Assured Survival*, p. 2. See footnote 3 in the Rand report done under Project Air Force (R-3369-AF).
28. Lawrence Freedman, *The Evolution of Nuclear Strategy*, p. 400.
29. See, for example, Maxwell Taylor, *The Uncertain Trumpet*, pp. 4-6, 30.
30. Michael Charlton, *The Star Wars History*, pp. 48, 49, 60. See Charlton's recounting of a BBC interview with Dr. Kissinger concerning issues the secretary raised in frustration in his famous question at a news conference in Moscow in June 1974: "What in the name of God is strategic superiority?" In his 1985 interview Kissinger said, "I do not think it is possible to achieve the degree of strategic superiority that existed in the 1940s and 1950s, such that we can rely on nuclear superiority to defend

Europe and other parts of the world." Yet in the same interview, Dr. Kissinger in a slightly different context and nuance to the term *superiority* said, "I have always believed, contrary to many scientists, that you're better off with superiority than with parity.... I do not think one can (or should) convince democratic publics that their extermination is the key element in their security, either the threat of their extermination or of *mutual* extermination. Even if superiority was not attainable, we were better off with a capacity for discriminating targeting than with a capacity for indiscriminate destruction...." On this issue of whether "superiority" has any meaning, Eugene Rostow in a BBC interview asserted that "the Soviets have a very clear answer [to Kissinger's question] ... what the Soviets are trying to do [in attempts to achieve superiority] is to assure American neutrality in the event of a war either in the Pacific or in the Atlantic."

31. Glenn A. Kent and Randall J. DeValk, *Strategic Defense and the Transition to Assured Survival*, p. v, note 2.

32. See Kent and DeValk for a comprehensive and brilliant elaboration of this argument (pp. 11-14, 44-45).

33. See the Kent and DeValk charts that lay out the "defense potential" of both Soviet and US forces in various combinations of offensive and defensive arsenals.

34. Keith B. Payne, *Strategic Defense*, pp. 96-101.

35. Ibid., p. 107.

36. Ibid. Payne elaborates these arguments, p. 107.

37. See Steven P. Adragna, *On Guard for Victory: Military Doctrine and Ballistic Missile Defense in the USSR* (Cambridge, Massachusetts: Pergamon-Brassey, 1987), p. 20 and pp. 22-27. Adragna quotes, on page 20, from a Soviet analyst (I.A. Grudinin) in a 1971 work: "Offense and defense constitute a dialectical unity of opposites, which simultaneously both exclude and assume one another. They are not only interconnected, but also mutually permeate one another and cannot exist separately." See also Ilana Kass and Ethan S. Burger, "Soviet Responses to the U.S. Strategic Defense Initiative: the ABM Gambit Revisited," *Air University Review* 30 (March-April 1985): 55-63.

38. Michael Charlton, *The Star Wars History*, p. 46. In a 1985 interview with BBC radio, Dr. Kissinger reflected on the action-reaction cycle in regard to MIRVs. "The Soviets started deploying MIRVs in 1973, which means that they were well along by 1970 and 1971, which is the time we may have had an opportunity [to decide not to go ahead]. As it turned out, we exploded

our H-bomb about nine months before the Soviets. We deployed our MIRVs about eighteen months before the Soviets. It is impossible therefore that they could have developed MIRVs just in that short interval between the time we did and they did."

39. See discussion in Colin S. Gray, "Strategic Defences: A Case for Strategic Defence," *Survival* 27 (March/April 1985): 50-55.

40. Ibid., p. 51.

5. POTENTIAL FOR A STABLE TRANSITION

1. Ambassador Paul H. Nitze, speech to the Philadelphia World Affairs Council, February 20, 1985. This speech was included in the USIS Wireless File, EUR 308, 02/20/85, pp. 18-23.

2. Ambassador Nitze used the same wording of the "central strategic concept" in both the Philadelphia World Affairs Council speech and in the 1985 Alastair Buchan Memorial Lecture that he delivered to the International Institute for Strategic Studies on March 28, 1985. The latter address was included in the USIS Wireless File, No. 60, March 29, 1985, pp. 2-15.

3. Ambassador Paul H. Nitze, World Affairs Council speech, p. 21. The phrase "mutual assured security" was an obvious contrast to the term "mutual assured destruction" (MAD), a phrase that has had currency from the early 1960s. The new term sought to capture the concept of deterrence based more and more on defensive stability and arms-race stability, "in which both countries possessed—and could continue to possess, regardless of an enemy's actions—the capability to survive as a nation under all circumstances." See also Glenn A. Kent and Randall J. DeValk, *Strategic Defenses and the Transition to Assured Survival*, p. vi.

4. Paul H. Nitze, "An Arms Control Agenda That Kissinger Should Know," *Washington Post*, March 30, 1987, Op-Ed page.

5. See Robert S. McNamara quoted in Michael Charlton, *The Star Wars History*, p. 23. McNamara went on to assert that even with the nuclear superiority of the United States at the time of the Cuban missile crisis, "we never conceived of using nuclear weapons under those circumstances. It was our tremendous conventional power in the region which forced the Soviets to take those missiles out."

6. Glenn A. Kent, *Strategic Defenses and the Transition to Assured Survival*, p. viii. The components of the US strategic

arsenal that do not receive adequate treatment in the RAND analysis are the potential of SLBMs (Trident missiles with D-5 warheads) and of cruise missiles (particularly SLCMs) to put large numbers of Soviet targets at risk, especially as the range, yield, and accuracy of the cruise missiles improve. Moreover, the results of any such analysis (based as it should be on current forces) can be altered radically through changes in assumptions about Pk ratios for Soviet and US ICBM warheads, about the capability of SLBMs to attack promptly "hardened" targets such as ICBM silos, the capability of layered strategic defenses to subtract out attacking warheads in a preferential approach, and the potential for arms control to reduce in an asymmetrical way the great advantage the Soviets now have in land-based ICBMs. At the same time, analysts certainly have to deal with what is and with what is planned, not with what might be and what would be "neat" to occur.

7. Ibid., p. vi.

8. One report that set out such steps was the April 1988 Defense Science Board study: See *Report of the Defense Science Board Task Force Subgroup on Strategic Air Defense (SDI Milestone Panel)*, Office of the Under Secretary of Defense for Acquisition (Washington, DC: US Government Printing Office, May 1988). See also General Accounting Office Report, "Strategic Defense System: Stable Design and Adequate Testing Must Precede Decision to Deploy" (Washington, DC: US Government Printing Office, July 1990).

9. See Marc S. Palevitz, "Beyond Deterrence: What the U.S. Should Do About Ballistic Missiles in the Third World," *Strategic Review* 18 (Summer 1990): 49-62.

10. See discussion in Michael Charlton, *The Star Wars History* (London: BBC Publications, 1986), pp. 11-12.

11. Compare conclusions in Glenn A. Kent, *Strategic Defenses and the Transition to Assured Survival*, pp. v-vii and pp. 44-45.

12. See National Conference of Catholic Bishops, "A Report on 'The Challenge of Peace' and Policy Development, 1983-1988," *Origins* 18 (1988): 133-48.

13. Andrew W. Marshall and Charles Wolf, *Sources of Change in the Future Security Environment*, The Pentagon, Washington DC, April 1988, p. 23. Marshall and Wolf were the working group chairmen for this report to the Commission on Integrated Long-Term Strategy. The commission in January 1988 had issued its report entitled *Discriminate Deterrence*.

14. Andrew W. Marshall and Charles Wolf, *Sources of Change in the Future Security Environment*, p. 9.

15. Strobe Talbott, "Inside Moves," *Time*, May 30, 1988, p. 31.

16. Ibid.

6. SDI'S BEARING ON NATO

1. H. W. Fowler, *A Dictionary of Modern English Usage* (Oxford: The Clarendon Press, 1926; reprint ed., 1934), pp. 558-60. See Fowler's essay on the "split infinitive" from which I adopted the categories.

2. See editorial by Hans A. Bethe and Glenn T. Seaborg, "Star Wars No, Nuclear Power Yes," *Washington Post*, September 16, 1988, A21.

3. "Report of the Defense Science Board Task Force Subgroup on Strategic Air Defense," prepared for the Office of the Under Secretary of Defense for Acquisition, May 1988. This report recommended, inter alia, that deployment "be in steps, each of which should provide some capability and have some value in itself." The first two deployment steps would be in compliance with the existing ABM Treaty limitations; the last four steps would require abrogation or modification of the Treaty.

See also Sam Cohen, "ALPS: First Mountain To Climb to Save SDI," *Wall Street Journal*, September 21, 1988, p. 26. Cohen then argued that President Reagan alone could save SDI from the presidency of either Mr. Dukakis (who called SDI a "fantasy and a fraud") or of Mr. Bush (who promised nothing about deployment). To do so, it was argued, Reagan had to accept the SDI version advocated by Senator Nunn: namely, the Accidental Launch Protection System (ALPS). ALPS is a limited ABM system "designed primarily to defend the U.S. against an accidental launch of intercontinental ballistic missiles." Cohen pegged the cost at $10 to $20 billion, and availability at within five years.

4. See Glenn Frankel, "Israel Puts Its First Satellite Into Orbit," *Washington Post*, September 20, 1988, p. A1.

5. Harald Muller, *Strategic Defences: The End of the Alliance Strategy* (Brussels, Belgium: Centre For European Policy Studies, 1987), p. 25.

Glossary

Anti-Ballistic Missile System—A missile system designed to intercept and destroy a strategic offensive ballistic missile or its reentry vehicles.

Anti-Satellite Weapon—A weapon designed to destroy satellites in space. The weapon may be launched from the ground or an aircraft or be based in space. The target may be destroyed by nuclear or conventional explosion, collision at high speed, or directed energy beam.

Architecture—Description of all functional activities to be performed to achieve the desired level of defense, the system elements needed to perform the functions, and the allocation of performance levels among those system elements.

Ballistic Missile—A guided vehicle propelled into space by rocket engines. Thrust is terminated at a predesignated time after which the missile's reentry vehicles are released and follow free-falling trajectories toward their ground targets under the influence of gravity. Much of a reentry vehicle's trajectory will be above the atmosphere.

Battle Management—A function that relies on management systems to direct target selection and fire control, and facilitate communications.

Boost—The first portion of a ballistic missile trajectory during which it is being powered by its engines. During this period, which usually lasts

---

Courtesy of the Strategic Defense Initiative Organization, 13 March 1989.

3 to 5 minutes for an ICBM, the missile reaches an altitude of about 200 km whereupon powered flight ends and the missile begins to dispense its reentry vehicles. The other portions of missile flight, including midcourse and reentry, take up the remainder of an ICBM's flight time of 25 to 30 minutes.

Booster—The rocket that propels the payload to accelerate it from the earth's surface into a ballistic trajectory, during which no additional force is applied to the payload.

Bus—Also referred to as a post-boost vehicle, it is the platform on which the warheads of a single missile are carried and from which warheads are dispensed.

Carrier Vehicle (CV)—A space platform whose principal function is to house the space-based interceptors in a protective environment prior to use.

Chaff—Strips of metal foil, wire, or metalized glass fiber used to reflect electromagnetic energy, usually dropped from aircraft or expelled from shells of rockets as a radar countermeasure.

Chemical Laser—A laser in which a chemical action is used to produce pulses of intense light.

Decoy—A device constructed to simulate a nuclear-weapon-carrying warhead. The replica is less costly and much less massive; it can be deployed in large numbers to complicate enemy efforts to read defense strategies.

Directed Energy—Energy in the form of atomic particles, pellets, or focused electromagnetic beams that can be sent long distances at, or nearly at, the speed of light.

Directed Energy Device—A device that employs a tightly focused and precisely directed beam of very intense energy, either in the form of light (a laser) or in the form of atomic particles traveling at velocities at or close to the speed of light (particle beams). (See also *Laser*.)

Discrimination—The process of observing a set of attacking objects and differentiating between decoys or other nonthreatening objects and actual threat objects.

Endoatmospheric—Within the earth's atmosphere, generally considered to be at altitudes below 100 kilometers.

Excimer Laser—Also called "excited dimer" laser, which uses the electrically produced excited states of certain molecules such as rare gas halides (which produce electromagnetic radiation in the visible and near ultraviolet part of the spectrum).

Exoatmospheric—Outside the earth's atmosphere, generally considered to be at altitudes above 100 kilometers.

Exoatmospheric Reentry Vehicle Interceptor Subsubsystem (ERIS)—The original name that refers to the Lockheed variant of a ground-based interceptor (GBI) that could be used in a strategic defense system.

Hardening—Measures which may be employed to render military assets less vulnerable.

Hypervelocity Gun (HVG)—A gun that can accelerate projectiles to 5 kilometers per second or more; for example, an electromagnetic or rail gun.

Imaging—The process of identifying an object by obtaining a high-quality image or profile of it.

Interception—The act of destroying a moving target.

Intercontinental Ballistic Missile (ICBM)—A land-based ballistic missile with a range greater than 3,000 nautical miles.

Intermediate-Range Ballistic Missile (IRBM)—A land-based ballistic missile with a range of 500 to 3,000 nautical miles.

Kinetic Energy—The energy from the motion of an object.

Kinetic Energy Interceptor—An interceptor that uses a nonexplosive projectile moving at very high speed to destroy a target on impact. The projectile may include homing sensors and on-board rockets to improve its accuracy, or it may follow a preset trajectory (as with a shell launched from a gun).

Laser (Light Amplification by the Stimulated Emission of Radiation)—A device for producing an intense beam of coherent light. The beam of light is amplified when photons (quanta of light) strike excited atoms or molecules. These atoms or molecules are thereby simulated to emit new photons (in a cascade or chain reaction) which have the same wavelength and are moving in phase and in the same direction as the original photon. A laser may destroy a target by heating, melting, or vaporizing its surface.

Layered Defense—A defense that consists of several layers that operate at different portions of the trajectory of a ballistic missile. Thus, there could be a first layer (e.g., boost) of defense with remaining targets passed on to succeeding layers (e.g., midcourse, terminal).

Leakage—The percentage of intact and operational warheads that get through a defensive system.

Lethality—State of effectiveness of an amount of energy or other beam characteristic required to eliminate the military usefulness of enemy targets by causing serious degradation or destruction of a target system.

Midcourse—That portion of the trajectory of a ballistic missile between boost/post-boost and reentry. During this portion of the missile trajectory, the target is no longer a single object but a swarm of RVs, decoys, and debris falling freely along preset trajectories in space.

Multiple Independently Targetable Reentry Vehicle (MIRV)— A package of two or more reentry vehicles which can be carried by a single ballistic missile and guided to separate targets. MIRVed missiles employ a warhead-dispensing mechanism called a post-boost vehicle which targets and releases the warheads.

Neutral Particle Beam (NPB)—An energetic beam of neutral atoms (no net electric charge). A particle accelerator accelerates the particles to nearly the speed of light.

Non-Nuclear Kill—Destruction that does not involve a nuclear detonation.

Particle Beam—A stream of atoms or subatomic particles (electrons, protons, or neutrons) accelerated to nearly the speed of light.

Particle Beam Device—A device that relies on the technology of particle accelerators (atom smashers) to emit beams of charged or neutral particles which travel near the speed of light. Such a beam could theoretically destroy a target by several means, e.g., electronics upsets, electronics damage, softening/melting of materials, sensor damage, and initiation of high explosives.

Passive Sensor—A sensor that detects only radiation naturally emitted (infrared radiation) or reflected (sunlight) from a target.

Penetration Aid—A device, or group of devices, that accompanies a reentry vehicle during its flight to spoof or misdirect defenses and thereby allow the RV to reach its target.

Post-Boost—The portion of a missile trajectory following boost and preceding midcourse.

Post-Boost Vehicle (PBV)—The portion of a missile payload that carries the multiple warheads and has maneuvering capability to place each warhead on its final trajectory to a target. (Also referred to as a "bus.")

Rail Gun—A device using electromagnetic launching to fire hypervelocity projectiles. Such projectile launchers will have very high muzzle velocities, thereby reducing the lead angle required to shoot down fast objects.

Reentry Vehicle (RV)—The part of a ballistic missile that carries the nuclear warhead to its target. The RV is designed to reenter the earth's atmosphere in the terminal portion of its trajectory and proceed to its target.

Sensor—A device that detects and/or measures certain types of physically observable phenomena.

Signature—The characteristic pattern of the target observed by detection and identification equipment.

Surveillance—An observation procedure that includes tactical observations, strategic warning, and meteorological assessments, by optical, infrared, radar, and radiometric sensors on spaceborne and terrestrial platforms.

Survivability—The capability of a system to avoid or withstand hostile environments without suffering irreversible impairment of its ability to accomplish its designated mission.

Terminal—The final portion of a ballistic missile trajectory during which warheads and penetration aids reenter the atmosphere. This follows midcourse and continues until impact or detonation.

Tracking and Pointing—Once a target is detected, it must be followed or "tracked." When the target is successfully tracked, an interceptor, laser, or neutral particle beam is "pointed" at the target. Tracking and pointing are frequently integrated operations.

INDEX

ABM Treaty. *See* Anti-Ballistic Missile (ABM) Treaty (1972)
Abrahamson, James A., 8, 38, 40–42, 91, 101
  as director of SDIO, 82, 185
  as proponent of SDI, 109
Abshire, David M., 24, 48, 101
Agnelli, Giovanni, 116
Anti-Ballistic Missile (ABM) Treaty (1972), 11
  concept of mutually assured destruction in, 22
  interpretation and revised interpretation of, 15–17, 22–23, 27, 38, 45–46, 49, 77, 124
  purpose of, 177
  Soviet violation of, 25, 123
  violation of, 13, 25
Anti-satellite weapons (ASAT), 98
  Soviet progress toward and possession of, 68, 113, 176–77
  US development of, 177
Anti-submarine warfare, 87
Anti-tactical ballistic missile (ATBM), 49, 51–53
  German interest in, 112
  NATO work on, 49, 51–53
Anti-tactical missile (ATM), 53, 186

Arms control, 7, 14, 27, 56–57, 59, 73, 96, 124, 125, 157, 161, 163, 171, 174, 180, 188, 190–92, 211–12
  negotiations for, 38–39, 161–62
Arms race, 154–59
Arnold, H. H., 3
ASAT. *See* Anti-satellite weapons (ASAT)
ATBM. *See* Anti-tactical ballistic missile (ATBM)
ATM. *See* Anti-tactical missile (ATM)

Ballistic missile defenses (BMD), 13, 132
Ballistic Missile Defense Systems Command, 41
Ballistic missiles
  defense against or elimination of, 13–14, 206
  possession of short- and intermediate-range, 176
  proliferation and sophistication of, 3, 13, 86–87, 205–6
Belgium, 93
Bethe, Hans, 109, 199
BMD. *See* Ballistic missile defenses (BMD)
Brilliant Pebbles concept, 91, 155–56
Brodie, Bernard, 5–6, 84
Brown, Harold, 186

247

248 INDEX

Bundy, McGeorge, 18, 128
Burden sharing, 155

Canada, 88, 93
Carrington, Peter, 16–17, 44
  management of SDI issues by, 76–78
Carter, Jimmy, 31, 125
Central Strategic Concept, 162–67
Chalfont, Alun, 99
Cheysson, Claude, 66, 106
China, 86
Chirac, Jacques, 20–21, 50, 89–90
Conventional Defense Initiative (CDI), 133
Courter, James, 55
Cruise missiles, 12
  as defense, 147
  defense against, 87
  See also Ground–launched cruise missile (GLCM)

Dam, Kenneth, 162–63
Deceptive basing schemes, 173
Defense Planning Committee (DPC), 51, 102
Defense Science Board Task Force Subgroup on Strategic Air Defense, 202–3
Defensive systems
  effectiveness of, 137
de Gaulle, Charles, 28, 103–4, 106, 193
DeLauer, Richard, 39
Denmark, 51, 88
Deterrence
  Ballistic missiles as, 100
  based on mutual assured destruction, 14
  in Central Strategic Concept, 164–67
  change in concept of, 5–6
  distinct from defense, 136–39
  European view of, 10–11
  morality of SDI, 18, 69–70, 114–15
  nuclear forces as, 98, 102, 135, 138
  role of second–strike forces for, 149–50
  shift with SDI of US strategy for, 12–13
  See also Stability
DeValk, Randall, 60, 145, 151, 172–73
Duerr, Hans-Peter, 109
Dukakis, Michael, 133
Dumas, Roland, 89
D–5 warheads, 12

East–West Study (1984), 44–45
Eureka program, 43, 68, 89–90, 106–7, 115
Euro-Group, 42, 102
European allies
  criticism of SDI by, 33–35
  deterrence position of, 10–11
  mistrust of US SDI plans, 81–83
  position on development of SDI, 12–16, 200–201
  reaction to SDI defense deployment, 36–37, 40
  reaction to SDI research proposal of, 37–38, 47–48

European Defense Initiative (EDI), 53, 99–100
Europeans
  perception of SDI protection of, 106
European Space Agency, 68
Extended Air Defense (EAD), 49, 53

Fletcher, James, 8, 11, 39–40, 52, 81, 198, 207
  as proponent of SDI, 109
Flexible response concept, 104–5, 126–29, 134–35, 139
Follow-on Forces Attack (FOFA) concept, 14, 114
Ford, Glyn, 66
France
  criticism of SDI by, 102–7
  Eureka program of, 43, 68
  participation in SDI of, 50, 88, 90–91
Freedman, Lawrence, 30–31, 48, 79, 136, 139–43, 145–46, 156–57
Fricaud-Chagnaud, Georges G., 135

Gallois, Pierre M., 135
Galosh interceptors, 68
Gang of Four, 18, 128
Garwin, Richard, 109, 199
Genscher, Hans Dietrich, 108–9
Germany, Federal Republic
  participation in Eureka program of, 89, 107–8, 115
  participation in SDI of, 50, 89, 93, 108

position on SDI of, 107–15
Glassboro meeting, 18, 35, 177
Glasser, Charles L., 80
GLCM. *See* Ground-launched cruise missile (GLCM)
Glenn, John, 43
Gorbachev, Mikhail, 14, 30, 59, 61, 115, 177
Graham, Daniel, 198
Gramm-Rudman-Hollings, 58
Gray, Colin S., 60, 156, 158
Greece, 51, 88
Ground-launched cruise missile (GLCM), 161
  intermediate-range, 114

Harmel Report (1967), 29, 44
Healey, Denis, 48, 97
Heseltine, Michael, 93
Hoffman, Fred S., 8, 11, 39–40, 133–34, 137, 142, 185–86
Homing Overlay Experiment, 41
Howe, Geoffrey, 44, 74–75, 98–99, 102
Huntington, Samuel, 136, 139

Ikle, Fred C., 137, 141–42
Independent European Programme Group, 42
Industrial conferences, 108
INF. *See* Intermediate-range nuclear forces (INF)
INF Treaty. *See* Intermediate-range

Nuclear Forces (INF) Treaty
Interceptors, 52, 87
Intercontinental ballistic missiles (ICBMs)
 argument for mobile, 173
 US and Soviet, 12, 19
Intermediate-range forces (INF)
 agreement on, 59
 deployment in Europe of, 38–40
Intermediate-range Nuclear Forces (INF) Treaty, 124
 effect of, 201–2
 missiles banned under, 132
 ratification of, 124
Iran-Iraq War, 13, 86, 176
Israel, 50, 92, 206
Italy
 participation in Eureka program of, 89
 participation in SDI of, 50, 89, 93, 115–19

Japan, 92–93
Jastrow, Robert, 81, 109, 166, 198, 207
JCS. *See* Joint Chiefs of Staff (JCS)
Johnson, Lyndon B., 35
Joint Chiefs of Staff (JCS), 60
Judd, O'Dean, 82

Kanter, Arnold, 60
Kemp, Jack, 55, 198
Kennan, George, 18, 128
Kent, Glenn, 60, 145, 151, 171, 172–73
Keyworth, George, 81, 198
Kirkpatrick, Jeane, 26
Kissinger, Henry, 28–29, 166
Kohl, Helmut, 20, 43, 109–14
Kosygin, Alexei, 18, 35, 177
Krasnoyarsk radar facility, 25, 41, 123
Krauthammer, Charles, 27–28, 52

Lambeth, Benjamin, 19–20
Lewis, Kevin, 19–20
Luxembourg, 50, 93

McFarlane, Robert, 141, 188
McNamara, Robert S., 18, 21, 28, 35, 177
 assured destruction position of, 140
 on flexible response concept, 126–27
 theory of deterrence of, 104–5
MAD. *See* Mutual assured destruction (MAD)
Menaul, Stuart B., 99–100
Miller, Franklin C., 8
Missile proliferation, 13
Mitterrand, Francois, 20, 50, 88, 89–91
Muller, Harald, 68
Multiple independently targetable reentry vehicles (MIRV), 147
Mutual assured destruction (MAD) concept, 6, 22, 34
 assured destruction, 128, 150, 152

effect of, 148–49
morality of, 14
MX missile, 12

NAC. *See* North Atlantic Council (NAC)
National Test Bed, 67
Netherlands, 70, 88, 93
Nitze, Paul, 42, 45, 78–79, 101
  criteria for SDI deployment of, 16–17
  development of strategic concept by, 162–67
  on interpretation of ABM Treaty, 22–23
  on morality of deterrence, 69–70
  on nuclear arms reduction, 162–67
Nixon, Richard M., 35
North Atlantic Council (NAC), 88, 102
North Atlantic Treaty Organization (NATO)
  Conventional Defense Improvements (CDI) Program of, 46–47, 83
  Defense Planning Committee (DPC) of, 88
  impact of SDI concept on strategy of, 28–30, 61–62
  plan for defense against ATBMs, 49
  reaction to SDI of, 30–32
  reaction to SDI research proposal, 36
  See also European allies
Norway, 50, 51
NPG. *See* Nuclear Planning Group (NPG)

Nuclear age
  new paradigm of war, 5–6
  responses to, 139–43
Nuclear Planning Group (NPG)
  ministerial meeting of (1984), 41
  ministerial meeting (1985), 9, 42, 87
  ministerial meeting (1986), 9–10, 53
Nuclear weapons
  effect on thinking of, 84
  European position on disarmament for, 158–59
  European position on efficacy of, 135–36
  proposal for reduction of, 161–67
  schools of thought in debate over, 180
  See also Central Strategic Concept
Nunn, Sam
  interpretation of ABM Treaty by, 15–16, 59
  position on SDI of, 57, 81, 186

Odom, William, 126
Office of Technology Assessment, 80, 186
Office of the Secretary of Defense (OSD)
  SDI office in, 40
Optical sensors, airborne, 52

Pakistan, 86
Parity, nuclear, 10
Payne, Keith, 60, 151, 152

Perle, Richard, 14, 42, 55–56, 74–75, 91, 101, 106
SDI exploitation position of, 188–89
Pershing IIs, 114, 161
Pike, John, 199
Portugal, 50, 93
Pressler, Larry, 109–10

Quayle, Dan, 198

Reagan, Ronald, 3, 12, 17, 20, 29, 30, 31, 33, 36
  advocates cooperative arrangements with Soviet Union, 175
  on defensive and offensive systems, 63–64
  desire to eliminate nuclear weapons, 33–34, 97, 142, 191
  impact on NATO of SDI concept, 61
  initial SDI speech of, 68–69
  position on elimination of ballistic missiles, 101
  position on transition to strategic defenses, 98
  SDI research proposal of, 37
  shift in position on SDI of, 140–41
  using SDI for assured security, 6–8
Reagan administration, 13–14, 55–56
Reykjavik summit, 30
  contradictory information about, 101
  on elimination of ballistic missiles, 14, 30, 49, 83, 101
Rogers, Bernard, 37, 65, 83
Roman Catholic Church, 69, 140, 186
Ruehle, Hans, 109

SALT. *See* Strategic Arms Limitations Treaty (SALT)
Saudi Arabia, 13, 86
Schlesinger, James, 31, 186
Schmidt, Helmut, 105
Scowcroft Commission (1983), 12, 83–84, 173
SDI Organization (SDIO), 8, 24–25, 40
SDI research
  parallels in ATBM defense initiative, 51
Sea-launched ballistic missiles (SLBMs), 97
Seitz, Fred, 81, 198
SHAPE, 60
Shultz, George P., 15, 46, 56
Single Integrated Operational Plan (SIOP), 31
SLBMs. *See* Sea-launched ballistic missiles (SLBMs)
Smith, Gerard, 18, 128
Soviet Union
  advantage in defenses of, 176–77
  anti-ballistic missile system of, 113
  ballistic missile system of, 8–11, 14
  ICBM warheads of, 151
  perception of defenses, 152–54

position on development of SDI, 18–20
potential cooperation in arms reduction with, 175
reaction to SDI research proposal, 36
strategic defense initiatives of, 61
violation of ABM Treaty by, 25

Space
defense weapons based in, 132
military use of, 62–63, 67–68, 119–20

Space systems, 154–55
Spain, 50, 93

Stability
of arms race, 154–59
defined, 145–49
first- and second-strike, 149–54
strategic, 143–45

Star Wars, 7–8
Strategic and non-strategic systems, 8–11
Strategic Arms Limitations Treaty (SALT), 8
Strategic Arms Reductions Talks (START), 59
Strategic concept, 8–11

Strategic Defense Initiative (SDI)
approval of research for, 20–21
concern of allies about, 33
country participation in, 49–51, 65–66, 87–94
criteria for deployment of, 16–17
critics and supporters of, 17–20, 48, 55, 65–67
deployment timetable of, 56–57, 58, 80–81
as deterrent, 29, 61–62, 69
European opposition to deployment of, 84
funding for, 46–47, 58, 92, 94, 124, 154–55, 202
goals and objectives of, 11–16, 39, 61–61, 156–59
impact of revised interpretation of ABM on, 46
misconceptions about, 24–25
opponents of, 18, 57, 109
perspectives on, 27–32
potential for strategic stability of, 143–45
proposed transition to technology of, 165–67
reaction to research proposal for, 36–38
research as hedge, 76
Soviet perception of, 18
studies of, 39–40, 52
suggested guidelines for United States, 70–71
supportive groups for, 65
See also European Defense Initiative (EDI)

Strategic defenses
convincing evidence for, 130–36
differing groups of observers of, 195–201
effect of unilateral deployment of, 177
position of limited defenses as, 185–87
position to exploit using, 187–90
proposal for non-nuclear, 162–67

as set of systems, 4–5, 184–85
time frame for deployment of, 26–27
See also Central Strategic Concept
Survivability issue, 21–23, 98, 142, 157–58

Tactical ballistic missiles (TBMs), 52
Talbott, Strobe, 123, 180
Taylor, Maxwell, 126
TBMs. *See* Tactical ballistic missiles (TBM)
Teller, Edward, 81, 109, 166, 198, 207
Thatcher, Margaret, 12, 20, 29, 59
  management of SDI issues by, 74–75, 94–98, 101
  persuasiveness of, 140–41
  role in arms control of, 57
Thatcher-Reagan four points agreement, 42–44, 96–98, 102, 140
Thomson, James, 60
Trident submarine, 12, 97, 98
Turkey, 50, 93
  Cesme, 41

Union of Concerned Scientists, 66–67, 80, 152, 199
United Kingdom
  participation in Eureka program, 89
  participation in SDI of, 49–50, 89, 92–93, 99
  Memorandum of Understanding, 92

Vulnerability issue, 21–23, 26, 83–84

Warnke, Paul, 174
Warsaw Pact, 136
Wehrkunde Conference (1986), 52
Weinberger, Caspar, 15, 41, 55–56, 88, 108–9, 166, 177
Western European Union (WEU), 42, 70, 93, 116
Woerner, Manfred, 9, 52–53, 109, 112
Wohlstetter, Albert, 60, 137, 141–42
Woolsey, James, 52

Younger, George, 102

ABOUT THE AUTHOR

Colonel Robert C. Hughes, USAF, is Associate Dean of Faculty and Academic Programs at the National War College, Washington, DC. Throughout his career, Colonel Hughes has had extensive experience in political-military affairs both on Air Staff and in assignments under the Joint Staff and the Office of the Secretary of Defense. After serving within Plans and Policy (AF/XOX) on the Air Staff in the late 1970s, he worked on special European projects in International Security Affairs (OSD/ISA) in the last year of the Carter administration. He then became Military Assistant to the Principal Deputy Assistant Secretary for International Security Policy (OASD/ISP) in the early 1980s. Colonel Hughes served as a member of the Chief's Staff Group when General Charles A. Gabriel, USAF, became Chief of Staff in the summer of 1982.

In mid-1984, Colonel Hughes joined the Defense Plans Division, United States Mission to NATO, Brussels, Belgium. From late 1984 through the summer of 1987, he had major responsibility under the Defense Adviser (Dr. Laurence J. Legere) and the Ambassador (The Honorable David M. Abshire) for issues and developments in NATO Europe related to the Strategic Defense Initiative and to anti-tactical missile efforts.

Colonel Hughes began this book in 1987-88 while assigned as a Senior Fellow at the National Defense University, where he graduated from the National War College. He earned a B. S. degree in philosophy at Our Lady of Providence College, an M.A. in linguistics and medieval studies from the

University of Rhode Island, and a Ph. D. in literature and linguistics from The Catholic University of America. Colonel Hughes also taught in the early 1970s at the Air Force Academy, Department of English, where he was an Associate Professor. In addition to his duties as Associate Dean at the National War College, Colonel Hughes teaches within the core curriculum and offers advanced studies courses, "National Space Strategy and Policy," as well as "NATO: Defense Perspectives."